Wireless Networks

Series Editor:
Xuemin (Sherman) Shen
University of Waterloo, Waterloo, Ontario, Canada

More information about this series at http://www.springer.com/series/14180

Yulong Zou • Jia Zhu

Physical-Layer Security for Cooperative Relay Networks

Yulong Zou
Nanjing University of Posts
    and Telecommunications
Nanjing, Jiangsu, China

Jia Zhu
Nanjing University of Posts
    and Telecommunications
Nanjing, Jiangsu, China

ISSN 2366-1186   ISSN 2366-1445   (electronic)
Wireless Networks
ISBN 978-3-319-80982-3   ISBN 978-3-319-31174-6   (eBook)
DOI 10.1007/978-3-319-31174-6

Printed on acid-free paper

This Springer imprint is published by Springer Nature
The registered company is Springer International Publishing AG Switzerland

Preface

Recently, cybercriminal activities in wireless communications systems are growing due to the fact that more and more emerging malware programs (also known as computer viruses) are targeted on the mobile terminals. Accordingly, an increasing attention has been paid to the research of wireless security against various malicious attacks. The radio propagation inherits the broadcast nature, leading to any receivers within the coverage area of a radio transmitter being capable of overhearing the wireless transmission. This makes the wireless communication systems extremely vulnerable to the eavesdropping attack. Typically, cryptographic techniques relying on secret keys are employed for preventing an eavesdropper from interpreting the wireless transmissions.

However, classic cryptographic techniques including both public-key cryptography and symmetric-key cryptography are only computationally secure and rely upon the hardness of their underlying mathematical problems. Cryptography security would be significantly compromised if an efficient method of solving its underlying mathematical problem was to be discovered. Moreover, conventional secret key exchange relies on a trusted key management center, which may not be always applicable in wireless networks. To this end, physical-layer security is emerging as a promising paradigm to secure wireless communications by exploiting the physical-layer characteristics of wireless channels. It is proved from an information-theoretic perspective that perfect secrecy can be achieved if the wiretap channel from a source to an eavesdropper is a degraded version of the main channel from the source to its intended destination. However, due to the time-varying fading effect of wireless channels, the main channel may experience a deep fade, which makes the perfect secrecy become impossible in some cases.

This book presents the concept and practical challenges of physical-layer security as well as examines recent advances in cooperative relaying for the wireless physical-layer security. In Chap. 1, we first review a range of physical-layer security techniques, namely, information-theoretic security, artificial-noise-aided security, security-oriented beamforming, and diversity-assisted security, along with an in-depth discussion of cooperative relaying techniques for wireless networks. Next, Chap. 2 investigates the physical-layer security for a wireless network consisting of a source and a destination with the aid of multiple relays, where only the single "best" relay is selected among the multiple relays to assist the source-destination transmission against eavesdropping. In Chap. 3, we then examine joint relay and jammer selection for enhancing the wireless physical-layer security of the source-destination transmission with the help of multiple intermediate nodes in the presence of an eavesdropper. In the joint relay and jammer selection, an intermediate node is selected to act as the relay for assisting the source-destination transmission and another intermediate node is chosen to act as the jammer for interfering with the eavesdropper.

Additionally, Chap. 4 explores the security-reliability tradeoff (SRT) for a wireless network, where security and reliability are measured by using the intercept probability experienced by the eavesdropper and outage probability encountered at the legitimate destination, respectively. We present two relay selection schemes for the SRT improvement, namely, single-relay selection (SRS) and multi-relay selection (MRS). To be specific, in the SRS scheme, only the single "best" relay is selected to assist the source-destination transmission, whereas in the MRS scheme, multiple relays are invoked to participate in forwarding the source signal to the destination. Finally, Chap. 5 re-examines the joint relay and jammer selection from the SRT perspective, where a relay is used to help the source transmission enhance wireless reliability and a friendly jammer is adopted to improve wireless security through the emission of the artificial noise for confusing the eavesdropper. It is shown that with an increasing number of relays and jammers, the security and reliability of wireless communications relying on the joint relay and jammer selection can be significantly enhanced concurrently.

Nanjing, China   Yulong Zou
January 2016   Jia Zhu

Contents

Acronyms

3G	Third generation
AF	Amplify-and-forward
AP	Access point
AWGN	Additive white Gaussian noise
BS	Base station
CDF	Cumulative distribution function
CF	Compress-and-forward
CRC	Cyclic redundancy check
CSI	Channel state information
DF	Decode-and-forward
DoS	Denial-of-service
GSVD	Generalized singular value decomposition
I.I.D	Independent identically distributed
LoS	Line-of-sight
LTE	Long-term evolution
LTE-A	Long-term evolution advanced
MAC	Medium access control
MER	Main-to-eavesdropper ratio
MIMO	Multiple-input multiple-output
MISOME	Multiple-input single-output multiple-eavesdropper
MRS	Multi-relay selection
OFDMA	Orthogonal frequency-division multiple access
PDF	Probability density function
SNR	Signal-to-noise ratio
SRS	Single-relay selection
SRT	Security-reliability tradeoff
TDMA	Time-division multiple access

Chapter 1
Introduction

Abstract Due to the broadcast nature of radio propagation, wireless transmissions
are accessible to any eavesdroppers and thus become extremely vulnerable to the
eavesdropping attack. Physical-layer security is emerging as a promising paradigm
to achieve the information-theoretic secrecy for wireless networks. This chapter
introduces the current research on physical-layer security for wireless networks. We
first discuss a range of physical-layer security techniques, including the information-
theoretic security, artificial noise aided security, security-oriented beamforming,
and diversity assisted security approaches. Then, we present an overview on
the cooperative relaying methods for wireless networks, namely the orthogonal
relaying, non-orthogonal relaying and relay selection. Additionally, the application
of cooperative relaying to wireless physical-layer security is also discussed for
protecting the wireless communications against eavesdropping.

1.1 Wireless Physical-Layer Security

Recent years have witnessed the widespread use of smartphones for accessing
various wireless networks, such as the third-generation (3G), long-term evolution
(LTE) and LET-advanced (LTE-A) mobile communications systems as well as
the Wi-Fi [1]. Meanwhile, it has been reported that an increasing number of
wireless terminals are compromised by the adversary for carrying out cybercriminal
activities, including malicious hacking, data forging, financial information theft,
and so on. Moreover, as discussed in [2] and [3], the broadcast nature of radio
propagation makes the wireless communication systems extremely vulnerable to
the eavesdropping attack. As shown in Fig. 1.1, an access point (AP) is considered
to transmit data packets to its associated legitimate users, which can be readily
overheard by an eavesdropper as long as it lies in the coverage area of AP.

Traditionally, cryptographic techniques relying on secret keys were adopted for
preventing an eavesdropper from interpreting the wireless data transmissions. There
are two main types of cryptographic techniques, namely the public-key cryptography and symmetric-key cryptography, which are however only computationally
secure and rely upon the hardness of their underlying mathematical problems [4, 5].
The security of a cryptographic approach would be significantly compromised, if
an efficient method of solving its underlying hard mathematical problem was to be

© Springer International Publishing Switzerland 2016   1
Y. Zou, J. Zhu, *Physical-Layer Security for Cooperative Relay Networks*,
Wireless Networks, DOI 10.1007/978-3-319-31174-6_1

Fig. 1.1 A wireless network consisting of an access point (AP) and multiple legitimate users in the face of an eavesdropper

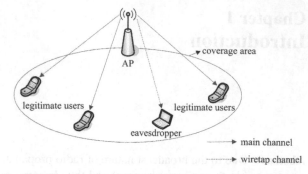

Fig. 1.2 Radio wave multipath propagation in an indoor environment

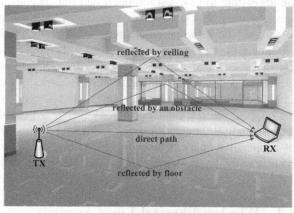

discovered [6]. Moreover, the conventional secret key exchange requires a trusted key management center, which may not be always applicable in wireless networks (e.g., wireless ad hoc networks).

As a consequence, physical-layer security emerges as an effective means to secure the wireless communications by exploiting the physical-layer characteristics of wireless channels [7–13]. It was proved in [7] that if the wiretap channel (from source to eavesdropper) is a degraded version of the main channel (from source to destination), the prefect secrecy can be achieved in an information-theoretic sense. In [8], the concept of secrecy capacity was further introduced, which is shown as the difference between the capacity of the main channel and that of the wiretap channel. More specifically, a positive secrecy capacity implies that the perfect secrecy is achievable and vice versa. However, due to the time-varying fading effect of wireless channels, the secrecy capacity of wireless communications is severely degraded, especially when a deep fade is encountered in the main channel between the source and destination.

As shown in Fig. 1.2, when a radio wave is emitted from the transmitter, it would be propagated over multiple different paths due to the radio reflection and scattering experienced by the obstacles as well as by the room ceiling and floor for indoor environments. This leads to the fact that multiple radiowave components arrive at the receiver, resulting in multipath reception. Since the differently delayed components sometimes add destructively, sometimes constructively, the strength of the radio signal combined at the receiver attenuates and fluctuates in time. The time-variant attenuation of multipath radio propagation is referred to as the *wireless fading*, which is modeled as a random process. There are three proper statistical models, namely the Rayleigh fading, Rician fading and Nakagami fading, which may be used for characterizing the wireless fading of radio propagation.

More specifically, Rayleigh fading is often used for modeling the magnitude attenuation of a radio wave signal propagated through a wireless medium (e.g., free space). This fading model is most applicable, when the propagation environment is comprised of massive obstacles. In such an environment, a radio wave signal would be heavily scattered and reflected, resulting in a large number of signal components arriving at the receiver. According to the central limit theorem, if the number of received radiowave components is sufficiently high, the combined signal containing the sine and cosine parts can be modeled as a complex-valued Gaussian random variable, regardless of the distribution of the individual components. The medium, which conveys the radio signal propagating through a wireless channel from the transmitter to receiver, is termed the channel response. To be specific, the real and imaginary parts of the channel response are modeled by independent and identically distributed (i.i.d) zero-mean Gaussian random variables. Hence, the amplitude of the channel response, denoted by x, is Rayleigh distributed with a probability density function (PDF) given by

$$p(x) = \frac{2x}{\sigma^2} \exp\left(-\frac{x^2}{\sigma^2}\right), x \geq 0 \tag{1.1}$$

where σ^2 represents an expected value of x^2 and can be viewed as the average power of the received radio signal. Again, the Rayleigh fading is very suitable for a propagation scenario in the presence of many scatterers to attenuate, reflect and diffract the radio signal.

Rician fading is another stochastic model commonly used for characterizing the multipath transmission when there is a line-of-sight (LOS) propagation path that has a higher signal strength than the other non-line-of-sight paths. By contrast, the above-mentioned Rayleigh fading is suitable when there is no LOS path. As mentioned above, in Rayleigh fading, the channel responses are assumed to be zero-mean Gaussian random variables. However, if there is a dominant LOS propagation path, then the mean of the random channel response should be around the power level of the dominant path, which is no longer zero. In such a situation containing a dominant LOS path, Rician fading is more appropriate than Rayleigh fading in modeling the transmission medium. In Rician fading, the channel amplitude x is modeled by a Rician distribution, which is expressed as

$$p(x) = \frac{2(K+1)x}{\sigma^2} \exp\left(-K - \frac{(K+1)x^2}{\sigma^2}\right) I_0\left(2\sqrt{\frac{K(K+1)}{\sigma^2}}x\right), x \geq 0 \tag{1.2}$$

where K is the ratio between the signal power of the direct LOS path and that of the other scattered paths, σ^2 is the average power of the total propagation paths, and $I_0(\cdot)$ is the first kind zero-order modified Bessel function. Additionally, when there is no direct LOS path, this means $K = 0$. It can be shown that substituting $K = 0$ into (1.2) readily leads to (1.1), implying that the Rayleigh distribution is a special case of the Rician distribution.

As discussed above, the multipath radio propagation is a random fading process, implying that a radio signal propagating over a wireless channel may be sometimes attenuated severely, sometimes only slightly. To this end, multiple antennas can be employed at a receiver so that multiple independently faded copies of the radio signal are received and combined together for the sake of enhancing the wireless reception quality. In this case, if the wireless channels between the transmitter and the receive antennas are modeled as Rayleigh fading, the amplitude of the combined signal containing multiple Rayleigh-faded components obey the Nakagami distribution associated with a shape factor m, where m is the number of receive antennas. Hence, Nakagami fading is used to model the amplitude of the sum of multiple i.i.d. Rayleigh-faded signals. The PDF of Nakagami distribution can be written as

$$p(x) = \frac{2m^m}{\Gamma(m)\sigma^{2m}}x^{2m-1}\exp\left(-\frac{m}{\sigma^2}x^2\right), x \geq 0 \tag{1.3}$$

where $\Gamma(m)$ represents the gamma function and σ^2 is the expectation of x^2. It has to be pointed out that when there is only a single receive antenna (i.e., $m = 1$), the Nakagami fading becomes the same as the Rayleigh fading, which can be readily validated by substituting $m = 1$ into (1.3) to obtain (1.1). The aforementioned three stochastic models, namely the Rayleigh fading, Rician fading and Nakagami fading, are generally utilized for characterizing the temporal behavior of a wireless propagation channel.

The wireless channel fading is assumed to remain constant during the coherence time by definition. However, in consecutive coherence time intervals, the wireless fading varies considerably and is modeled as a stochastic process. By comparing the coherence time of a wireless channel and the symbol period of a radio signal to be transmitted over the channel, the fading imposed by the channel may be classified as slow fading and fast fading. More specifically, slow fading is encountered when the coherence time is higher than the symbol period, where again, the fading amplitude and phase are considered to be constant over the symbol period. By contrast, fast fading implies that the coherence time is shorter than the symbol period, where the wireless fading varies during the symbol period.

Additionally, in a wireless channel, the coherence bandwidth represents the range of frequencies, for which a radio signal transmitted over the channel will experience comparable fading attenuation across the different frequencies. If the coherence bandwidth of the channel is higher than the frequency bandwidth of the radio signal, then all frequency components of the signal are deemed to experience more or less the same fading effect, which is referred to as flat fading. On the other hand, when the coherence bandwidth becomes lower than the bandwidth of the signal, different frequency components of the signal will experience different fading effects. This is referred to as frequency-selective fading. Therefore, the radio propagation may also be divided into frequency-flat and frequency-selective fading.

In order to combat the fading effect for enhancing the wireless secrecy capacity, considerable research efforts were devoted to the development of various physical-layer security techniques, which can be classified into the following main research categories: (1) the information-theoretic security [10–16], (2) artificial noise aided security [17–21], (3) security-oriented beamforming [22–26], and (4) diversity assisted security approaches [27–31].

Information-theoretic security is to examine the fundamental limits of physical-layer security from information-theoretic perspective. Historically, the information-theoretic security was first studied by Shannon in [10], with an emphasis on the mathematical properties of communications secrecy systems. More specifically, a secrecy system was achieved by mathematically transforming the plaintext into the cryptograms with the aid of secret keys, where the transformation shall be nonsingular so that unique deciphering is possible given the secret key. In [10], the theoretical secrecy was developed to address the communications security against an eavesdropper either with infinite or finite computing power. It was shown in [10] that a perfect secrecy system is achievable with a finite secret key, where an eavesdropper is unable to obtain a unique deciphering solution. In [7], Wyner investigated the information-theoretic security for a discrete memoryless wiretap channel consisting of a source, a destination and an eavesdropper. It was proved in [11] that when the main channel has a better condition than the wiretap channel, there exists a positive rate at which the source and destination can exchange information at perfect secrecy. In [8], Wyner's results were extended to the Gaussian wiretap channel, where the secrecy capacity is shown as the difference between the capacity of the main channel and that of the wiretap channel.

For the sake of improving the wireless secrecy in fading environments, the multiple-input multiple-output (MIMO) was studied extensively to combat the wireless fading and increase the secrecy capacity. In [14], Khisti and Wornell investigated a multiple-input single-output multiple-eavesdropper (MISOME) sce-nario as shown in Fig. 1.3, where the source and eavesdropper are assumed with multiple antennas and the destination has only one antenna. By assuming that the eavesdropper's channel state information (CSI) is known, the secrecy capacity of the MISOME was characterized in terms of the generalized eigenvalues. When the CSI knowledge of the eavesdropper becomes unavailable, Khisti and Wornell further presented a masked beamforming scheme and showed that it can obtain a near-optimal security performance in high signal-to-noise ratio (SNR) regions. In [15], the authors examined the information-theoretic security from the MISOME case to a more general MIMO scenario compromised of the source, destination and eavesdropper each equipped with multiple antennas. They considered two specific cases: (1) the deterministic case in which the CSIs of both the man channel and wiretap channel are fixed and known; and (2) the fading case in which both the main channel and wiretap channel experience Rayleigh and only the statistical CSI of the wiretap channel is assumed. The generalized-singular-value-decomposition

Fig. 1.3 A multiple-input single-output multiple-eavesdropper (MISOME) communications system

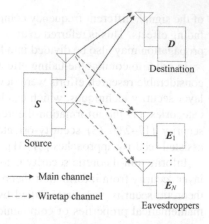

> → Main channel
> --→ Wiretap channel

(GSVD) approach was utilized to achieve the secrecy capacity for the deterministic case, which was then extended to the fading case, showing that the corresponding secrecy capacity approaches to zero, if the ratio of the number of eavesdropper's antennas to that of the source's antennas is larger than two.

The artificial noise aided security [17–21] enables the source to generate an interfering signal (referred to as artificial noise) in a manner such that the artificial noise only interferes with the eavesdropper, while the legitimate destination keeps unaffected. This results in a decrease of the capacity of the wiretap channel without affecting the capacity of the main channel, thus leading to an increase of the secrecy capacity. In [17], the authors considered a wireless network comprised of a source and a destination in the face of an eavesdropper and investigated the use of the artificial noise for defending the source-destination transmission against eavesdropping. More specifically, a certain transmit power at the source is allocated to generate the artificial noise, which is sophisticatedly designed such that only the wiretap channel is degraded without affecting the main channel. To this end, the authors of [17] proposed the employment of multiple antennas to generate the artificial noise, and proved that the artificial noise would not degrade the main channel only when the number of transmit antennas at the source is more than the number of receive antennas at the eavesdropper. It was shown in [17] that a positive secrecy capacity can always be achieved by using the artificial noise, even if the wiretap channel is better than the main channel. Although the artificial noise aided security is able to guarantee the secrecy of wireless communications, it comes at the expense of additional power consumption since a certain transmit power is needed for producing the artificial noise. In [18], the authors studied the power allocation between the information-bearing signal (that carries the desired information from source to destination) and the artificial noise (that is employed to interfere with the eavesdropper). It was shown in [18] that the simple equal power allocation is a near-optimal strategy when the eavesdroppers are uncoordinated and independent of each other. In addition, when the CSI estimation error is considered, it was observed that using more power for the artificial noise can achieve better secrecy performance than increasing the transmit power for the information-bearing signal.

The security-oriented beamforming [22–26] allows the source to transmit its signal in a particular direction to the destination such that the signal arrived at an eavesdropper experiences destructive interference and becomes much weaker than that received at the destination experiencing constructive interference. In this way, the capacity of the main channel from the source to destination relying on the security-oriented beamforming would be much higher than that of the wiretap channel from the source to eavesdropper, leading to the enhancement of secrecy capacity. In [22], the authors examined the use of relay nodes to form a beamforming system with the perfect CSI knowledge of the main channel and wiretap channel. The relay beamforming design was performed to maximize the secrecy rate under the total transmit power constraint, which was addressed by using the well-known semi-definite programming technique. It was shown in [22] that the proposed beamforming method significantly increases the secrecy capacity of wireless communications. In [23], multiple antennas were utilized for the beamforming design to improve the secrecy capacity of the source-destination transmission in the presence of an eavesdropper. Differing from the beamforming design of [22] relying on the perfect CSI knowledge of the wiretap channel, Mukherjee et al. [23] carried out the beamforming design without knowing the eavesdropper's CSI knowledge. In addition, the authors of [24] investigated the transmit beamforming in a wireless relay network consisting of a source, a destination and an untrusted relay, where the relay may potentially assist the signal transmission from source to destination. But, the relay is untrusted in the sense that it may become a passive eavesdropping attack. Two secure beamforming schemes, namely the noncooperative beamforming and cooperative beamforming, were proposed to maximize the wireless secrecy capacity. Simulation results showed that both the noncooperative and cooperative beamforming schemes performs better than the conventional approaches in terms of the secrecy capacity.

As an alternative, diversity techniques [27–31] can also be utilized to enhance the wireless security against eavesdropping attack, called the diversity assisted security. For example, considering that the source has multiple transmit antennas, an optimal antenna can be selected and used to transmit the desired signal depending on whether the CSI of the main channel and wiretap channel is available [27]. To be specific, if the CSI of both the main channel and wiretap channel is known at the source, a transmit antenna with the highest instantaneous secrecy capacity can be chosen to transmit the desired signal, which can significantly improve the secrecy capacity of wireless transmission. If only the main channel's CSI is available, we can choose a transmit antenna with the highest instantaneous capacity of the main channel to send the desired signal. Since the transmit antenna selection is based on the main channel's CSI only and the wiretap channel is typically independent of the main channel, the capacity of the main channel will be increased with the transmit antenna selection, but no capacity improvement can be achieved for the wiretap channel. This finally results in an increase of the secrecy capacity through the transmit antenna selection.

In addition, the multiuser diversity [30, 31] is another effective means of improving the wireless physical-layer security against eavesdropping. As shown

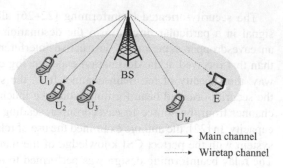

Fig. 1.4 A cellular network consisting of a base station (BS) and multiple users in the presence of an eavesdropper

—————▸ Main channel
- - - - - -▸ Wiretap channel

in Fig. 1.4, we consider a cellular network comprised of a base station (BS) and multiple associated users. Typically, a multiple access approach, such as the time-division multiple access (TDMA) and orthogonal frequency-division multiple access (OFDMA) is employed to enable the multiple users to communicate with the BS. Taking the TDMA as an example, we should first perform the multiuser scheduling to determine which user is scheduled to access a given time slot for data transmissions. In order to effectively protect the wireless transmissions against eavesdropping, the multiuser scheduling should be aimed at minimizing the capacity of the wiretap channel, while maximizing the capacity of the main channel. If the CSIs of both the main channel and the wiretap channel are known, a specific user with the highest instantaneous secrecy capacity would be chosen to access the given time slot. If only the main channel's CSI is known, then a user that maximizes the capacity of the main channel is selected for accessing the slot, which does not need the wiretap channel's CSI.

1.2 Overview of Cooperative Relay Networks

This section presents an overview of cooperative relay techniques for wireless networks. In general, there are two different types of relays, namely the full-duplex relay and the half-duplex relay [32, 33]. More specifically, the full-duplex relay implies that a relay can transmit and receive radio signals simultaneously over the same channel. By contrast, with the half-duplex relay, two orthogonal channels are needed for the relay to transmit and receive the radio signals. It is obvious that the full-duplex relay saves the spectrum resource and doubles the spectrum utilization, as compared to the half-duplex relay. However, with the full-duplex relay, an incoming signal received at an antenna will be interfered with its outgoing signal, which is known self-interference. It is challenging to cancel out such a self-interference due to the significant difference in the power levels of incoming and outgoing signals [34]. Thus, although the half-duplex relay has a lower spectrum utilization than the full-duplex relay, it is still used in practical wireless systems due to its simplicity in implementation. To alleviate the loss of spectrum utilization, the two-way relay was examined by employing the physical-layer network coding [35, 36], where the exchange of two messages can be completed between two source nodes via the relay over two orthogonal channels.

As shown in Fig. 1.5, when a relay node assists the message transmission from a source to a destination, it typically needs two phases [37–39]: (1) multicast phase, where the source multicasts its signal to the relay and destination, and (2) retransmission phase, where the relay forwards its received signal to the destination by using a certain relaying protocol, i.e., the amplify-and-forward (AF) protocol and decode-and-forward (DF) protocol. Figure 1.6 shows a comparison between the two AF and DF protocols. It can be seen from Fig. 1.6 that with the AF protocol, the relay node just simply retransmits a scaled version of its received noisy signal to the destination. By contrast, the DF protocol enables the relay first to decode its received signal and then to forward its decoded result to the destination, as shown in Fig. 1.6.

More specifically, the AF is a simple relaying protocol, which allows a relay node to amplify and retransmit its received noisy version of the source signal

Fig. 1.5 A wireless network consisting of a source, a relay and a destination

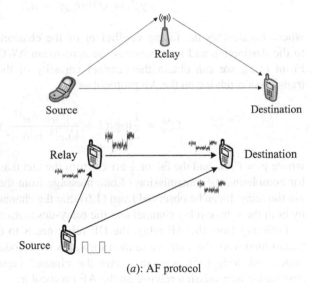

Source   Destination

Fig. 1.6 Comparison between the amplify-and-forward (AF) and decode-and-forward (DF) protocols: (**a**) AF relaying and (**b**) DF relaying

(*a*): AF protocol

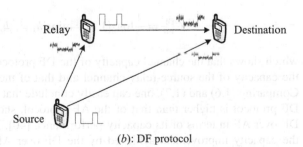

(*b*): DF protocol

to the destination. On the one hand, the AF protocol has the advantage of simple implementation, since the AF relay just forwards its received noisy signal without relying on any kind of decoding operations. On the other hand, the main disadvantage of the AF protocol is that the noise received at the relay node may also be amplified and forwarded to the destination. This would result in a performance degradation at the destination in decoding the source signal. Considering that the source transmits its signal denoted by x at a power of P, we can express the received signal at the relay as

$$y_r = \sqrt{P}h_{sr}x + n_r, \tag{1.4}$$

where h_{sr} denotes the fading coefficient of the channel spanning from the source to the relay and n_r represents the zero-mean additive white Gaussian noise (AWGN) with a variance of N_0. Next, the AF relay amplifies its received signal y_r with a scaling factor $\alpha = 1/(\sqrt{P}|h_{sr}|)$ and retransmits the scaled signal to the destination. Thus, the received signal at the destination can be written as

$$y_d^{AF} = \sqrt{P}\alpha h_{rd}y_r + n_d, \tag{1.5}$$

where h_{rd} denotes the fading coefficient of the channel spanning from the relay to the destination and n_d represents the zero-mean AWGN with a variance of N_0. From (1.5), we can obtain the channel capacity of the source-relay-destination transmission relying on the AF protocol as

$$C_{srd}^{AF} = \frac{1}{2}\log_2(1 + \frac{|h_{sr}|^2|h_{rd}|^2}{|h_{rd}|^2 + |h_{sr}|^2}\gamma), \tag{1.6}$$

where $\gamma = P/N_0$ and the factor $\frac{1}{2}$ arises from the fact that two time slots are needed for completing the transmission of one message from the source to the destination via the relay. It can be observed from (1.6) that the channel capacity C_{srd}^{AF} is affected by both the source-relay channel and the relay-destination channel.

Differing from the AF relay, the DF relay needs to decode its received signal transmitted from the source and then forwards its decoded result to the destination node. Following [37], we can obtain the channel capacity of the source-relay-destination transmission relying on the AF protocol as

$$C_{srd}^{DF} = \frac{1}{2}\log_2[1 + \gamma \min(|h_{sr}|^2, |h_{rd}|^2)], \tag{1.7}$$

which shows that the channel capacity of the DF protocol is the minimum between the capacity of the source-relay channel and that of the relay-destination channel. Comparing (1.6) and (1.7), one can easily conclude that the channel capacity of the DF protocol is higher than that of the AF protocol, showing the advantage of the DF over AF in terms of its capacity performance [40]. It is worth mentioning that the capacity improvement achieved by the DF over AF comes at the expense of

Fig. 1.7 A source transmits
to its destination with the aid
of multiple relays

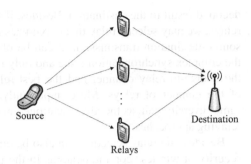

Source

Relays

Destination

communications latency, since the decoding operations required by the DF protocol incur an additional time delay, whereas the AF just simply forwards its received signal without any sort of decoding.

So far, we have discussed two basic relaying protocols, namely the AF and DF for a single-relay wireless network. As shown in Fig. 1.7, multiple relays may become available to assist the transmission from the source to destination. Given multiple relay nodes available, it is of particular interest to explore how the relays can be utilized in an efficient manner so that a significant performance enhancement is achieved while the corresponding cost is minimized. When multiple relays become available, we may allow all the relays to forward their received signals from the source to the destination in two different ways, namely the orthogonal relaying and non-orthogonal relaying [41–43]. More specifically, in the orthogonal relaying process, all the relays retransmit their received source signals over mutually orthogonal channels. This can effectively avoid the signal interference among different relays, but scarifies the precious channel resources, since an increasing orthogonal channels are required as the number of relays increases. By contrast, the non-orthogonal relaying enables the multiple relays to transmit simultaneously over the same channel, which significantly saves the orthogonal channel resources. However, it requires complex symbol-level synchronization among multiple spatially distributed relays so that the different signals transmitted by different relays can be synchronized and combined at the destination without causing interference. Moreover, as the number of relays increases, the complexity of such symbol-level synchronization becomes extremely high.

Alternatively, the relay selection was also studied to address the multiple-relay scenario, where only the single "best" relay is chosen to participate in forwarding the source signal to the destination [44–46]. To be specific, the source first broadcasts its message to all relay nodes, which attempt to decode the source message from their received signals, if the DF protocol was used at the relays. Let these relays which successfully decode the source message constitute a so-called *decoding set*. It needs to be pointed out that the cyclic redundancy check (CRC) coding may be invoked for determining whether a relay node succeeds in decoding its received signal or not. Particularly, if a relay nodes passes the CRC checking, it is considered to be successful in decoding and added to the decoding set. If the decoding set is not empty, a single relay would be chosen among the decoding set for forwarding its decoded result to the destination. Besides, if the AF protocol was considered at the relays, we may select a relay that maximizes the received SNR y_d^{AF} for assisting the source-destination transmission. It can be observed that the relay selection avoids the complex synchronization issue and only requires two orthogonal channels (i.e., the source-to-relay channel and the best-relay-to-destination channel) regardless of the number of relays. More importantly, the relay selection can achieve the same diversity gain as the aforementioned non-orthogonal relaying and orthogonal relaying approaches.

Besides, the relay selection can also be employed to enhance the physical-layer security of wireless communications in the presence of an eavesdropper [47, 48]. More specifically, considering a wireless network consisting of a source and a destination with the help of multiple relay nodes, we can choose the "best" relay to assist the source-destination transmission against eavesdropping. To elaborate a litter further, if the CSIs of both the main channel and wiretap channel are available, a relay node with the highest instantaneous secrecy capacity may be invoked to forward the source signal to the legitimate destination. If the eavesdropper's CSI is unknown and only the main channel's CSI is available, then a relay node with the highest instantaneous capacity of the main channel is selected to assist the source transmission to the destination. It was shown in [47] that the wireless secrecy capacity can be significantly improved relying on the relay selection, which explicitly demonstrates the physical-layer security benefits of exploiting the cooperative relays against eavesdropping.

1.3 Objective of This Book

Cryptographic techniques relying on secret keys are typically employed in order to guarantee the transmission confidentiality against eavesdropping, which are only computationally secure and depend on the hardness of their underlying mathematical problems. Moreover, the key distribution relies upon a trusted infrastructure, which may be unavailable and even compromised in some cases. To this end, the emerging physical-layer security exploits physical characteristics of wireless channels for achieving the perfect secrecy against eavesdropping. However, the physical-layer security of wireless transmissions is severely limited and degraded by the time-varying multipath fading.

The objective of this book is to present the concept and practical challenges of physical-layer security as well as examine recent advances in cooperative relaying designs for the wireless security enhancement. The motivations and concepts of physical-layer security are first explored along with a review of the wireless security threats in cooperative relay networks. Then, the relay-selection designs are discussed in details for improving wireless secrecy against eavesdropping in time-varying fading environments. Finally, the security-reliability tradeoff (SRT) is mathematically characterized for wireless communications. Also, the relay selection as well as the joint relay and jammer selection are examined for the wireless SRT improvement.

References

1. O. Aliu, A. Imran, M. Imran, and B. Evans, "A Survey of self organisation in future cellular networks," *IEEE Commun. Surveys & Tutorials*, vol. 15, no. 1, pp. 336–361, Feb. 2013.
2. A. Mukherjee, S. A. Fakoorian, J. Huang, and A. L. Swindlehurst, "Principles of physical layer security in multiuser wireless networks: A survey," *IEEE Commun. Surveys & Tutorials*, vol. 16, no. 3, pp. 1550–1573, Sept. 2014.
3. Y. Zou, J. Zhu, L. Yang, Y.-.C. Liang, and Y.-D. Yao, "Securing physical-layer communications for cognitive radio networks," *IEEE Commun. Mag.*, vol. 53, no. 9, pp. 48–54, Sept. 2015.
4. S. Mathur, W. Trappe, N. Mandayam, C. Ye, and A. Reznik, "Radio-telepathy: Extracting a secret key from an unauthenticated wireless channel," in *Proceedings of The 14th Annual International Conference on Mobile Computing and Networking (MobiCom 2008)*, California, USA, September 2008.
5. S. Jana, *et al.*, "On the effectiveness of secret key extraction from wireless signal strength in real environments," in *Proceedings of The 15th Annual International Conference on Mobile Computing and Networking (MobiCom 2009)*, Beijing, China, September 2009.
6. Q. Wang, K. Xu, and K. Ren, "Cooperative secret key generation from phase estimation in narrowband fading channels," *IEEE J. Sel. Areas Commun.*, vol. 30, no. 9, pp. 1666–1674, Sept. 2012.
7. A. D. Wyner, "The wire-tap channel," *Bell Syst. Tech. J.*, vol. 54, no. 8, pp. 1355–1387, 1975.
8. S. K. Leung-Yan-Cheong and M. E. Hellman, "The Gaussian wiretap channel," *IEEE Trans. Inf. Theory*, vol. 24, no. 7, pp. 451–456, Jul. 1978.
9. Y. Zou, X. Li, and Y.-C. Liang, "Secrecy outage and diversity analysis of cognitive radio systems," *IEEE J. Sel. Areas Commun.*, vol. 32, no. 11, pp. 2222–2236, Nov. 2014.
10. C. E. Shannon, "Communications theory of secrecy systems," *Bell Syst. Tech. J.*, vol. 28, pp. 656–715, 1949.
11. H. Mahdavifar and A. Vardy, "Achieving the secrecy capacity of wiretap channels using polar codes," *IEEE Trans. Inf. Theory*, vol. 57, no. 10, pp. 6428–6443, Oct. 2011.
12. P. K. Gopala, L. Lai, and H. E. Gamal, "On the secrecy capacity of fading channels," *IEEE Trans. Inf. Theory*, vol. 54, no. 10, pp. 4687–4698, Oct. 2008.
13. Y. Liang, H. V. Poor, and S. Shamai, "Secure communication over fading channels," *IEEE Trans. Inf. Theory*, vol. 54, no. 6, pp. 2470–2492, Jun. 2008.
14. A. Khisti and G. W. Wornell, "Secure transmission with multiple antennas: The MISOME wiretap channel," *IEEE Trans. Inf. Theory*, vol. 56, no. 7, pp. 3088–3104, Jul. 2010.
15. F. Oggier and B. Hassibi, "The secrecy capacity of the MIMO wiretap channel," *IEEE Trans. Inf. Theory*, vol. 57, no. 8, pp. 4961–4972, Aug. 2011.
16. S. Shafiee, N. Liu, and S. Ulukus, "Towards the secrecy capacity of the Gaussian MIMO wiretap channel: The 2-2-1 channel," *IEEE Trans. Inf. Theory*, vol. 55, no. 9, pp. 4033–4039, Sept. 2009.
17. S. Goel and R. Negi, "Guaranteeing secrecy using artificial noise," *IEEE Trans. Wirel. Commun.*, vol. 7, no. 6, pp. 2180–2189, Jul. 2008.
18. X. Zhou and M. McKay, "Secure transmission with artificial noise over fading channels: Achievable rate and optimal power allocation," *IEEE Trans. Veh. Tech.*, vol. 59, no. 8, pp. 3831–3842, Aug. 2010.
19. D. Goeckel, *et al.*, "Artificial noise generation from cooperative relays for everlasting secrecy in two-hop wireless networks," *IEEE J. Sel. Areas Commun.*, vol. 29, no. 10, pp. 2067 2076, Oct. 2011.
20. S.-C. Lin, *et al.*, "On the impact of quantized channel feedback in guaranteeing secrecy with artificial noise: The noise leakage problem," *IEEE Trans. Wirel. Commun.*, vol. 10, no. 3, pp. 901–915, Mar. 2011.
21. Q. Li and W.-K. Ma, "Spatially selective artificial-noise aided transmit optimization for MISO multi-eves secrecy rate maximization," *IEEE Trans. Signal Process.*, vol. 61, no. 10, pp. 2704–2717, Oct. 2013.

22. J. Zhang and M. Gursoy, "Collaborative relay beamforming for secrecy," in *Proc. The 2010 IEEE Intern. Conf. Commun.*, Cape Town, South Africa, May 2010.
23. A. Mukherjee amd A. Swindlehurst, "Robust beamforming for security in MIMO wiretap channels with imperfect CSI," *IEEE Trans. Signal Process.*, vol. 59, no. 1, pp. 351–361, Jan. 2011.
24. C. Jeong, I. Kim, and K. Dong, "Joint secure beamforming design at the source and the relay for an amplify-and-forward MIMO untrusted relay system," *IEEE Trans. Signal Process.*, vol. 60, no. 1, pp. 310–325, Jan. 2012.
25. Y. Yang, *et al.*, "Cooperative secure beamforming for AF relay networks with multiple eavesdroppers," *IEEE Sig. Process. Lett.*, vol. 20, no. 1, pp. 35–38, Jan. 2013.
26. X. Liu, *et al.*, "Joint beamforming and user selection in multicast downlink channel under secrecy-outage constraint," *IEEE Commun. Lett.*, vol. 18, no. 1, pp. 82–85, Jan. 2014.
27. Y. Zou, J. Zhu, X. Wang, and V. Leung, "Improving physical-layer security in wireless communications through diversity techniques," *IEEE Netw.*, vol. 29, no. 1, pp. 42–48, Jan. 2015.
28. Y. Zou, B. Champagne, W.-P. Zhu, and L. Hanzo, "Relay-selection improves the security-reliability trade-off in cognitive radio systems," *IEEE Trans. Commun.*, vol. 63, no. 1, pp. 215–228, Jan. 2015.
29. A. Jindal, C. Kundu, and R. Bose, "Secrecy outage of dual-hop AF relay system with relay selection without eavesdropper's CSI," *IEEE Commun. Lett.*, vol. 18, no. 10, pp. 1759–1762, Oct. 2014.
30. Y. Zou, X. Wang, and W. Shen, "Physical-layer security with multiuser scheduling in cognitive radio networks," *IEEE Trans. Commun.*, vol. 61, no. 12, pp. 5103–5113, Dec. 2013.
31. X. Ge, *et al.*, "Secrecy analysis of multiuser downlink wiretap networks with opportunistic scheduling," in *Proc. 2015 Intern. Conf. Commun.*, London UK, Jun. 2015.
32. B. Rankov and A. Wittneben, "Spectral efficient protocols for half-duplex fading relay channels," *IEEE J. Sel. Areas Commun.*, vol. 25, no. 2, pp. 379–389. Feb. 2007.
33. I. Krikidis, *et al.*, "Full-duplex relay selection for amplify-and-forward cooperative networks," *IEEE Trans. Wirel. Commun.*, vol. 11, no. 12, pp. 4381–4393, Dec. 2012.
34. S. W. Peters, A. Y. Panah, K. T. Truong, and R. W. Heath, "Relay architectures for 3GPP LTE-advanced," *EURASIP J. Wireless Commun. and Net.*, Vol. 2009, doi:10.1155/2009/618787.
35. S. Zhang, S. C. Liew, and P. P. Lam, "Hot topic: Physical-layer network coding," in *Proc. The 12th ACM Annual Intern. Conf. Mobile Comp. Net.*, Los Angeles, CA, Sept. 2006.
36. R. Louie, Y. Li, and B. Vucetic, "Practical physical layer network coding for two-way relay channels: Performance analysis and comparison," *IEEE Trans. Wirel. Commun.*, vol. 9, no. 2, pp. 764–777, Feb. 2010.
37. J.N. Laneman, D.N.C. Tse, and G.W. Wornell, "Cooperative diversity in wireless networks: Efficient protocols and outage behavior," *IEEE Trans. Inf. Theory*, vol. 50, no. 12, pp. 3062–3080, Dec. 2004.
38. A. Sendonaris, E. Erkip, and B. Aazhang, "User cooperation diversity part I: System description," *IEEE Trans. commun.*, vol. 51, no. 11, pp. 1927–1938, Nov. 2003.
39. Y. Zou, Y.-D. Yao, and B. Zheng, "Opportunistic distributed space-time coding for decode-and-forward cooperation systems," *IEEE Trans' Signal Process.*, vol. 60, no. 4, pp. 1766–1781, Apr. 2012.
40. M. Souryal and B. Vojcic, "Performance of amplify-and-forward and decode-and-forward relaying in Rayleigh fading with turbo codes," in *Proc. 2006 IEEE Intern. Conf. Acoustics, Speech and Signal Process. (IEEE ICASSP 2006)*, Toulouse, France, May 2006.
41. J. Laneman and G. W. Wornell, "Distributed space-time-coded protocols for exploiting cooperative diversity in wireless networks," *IEEE Trans. Inf. Theory*, vol. 49, no. 10, pp. 2415–2425, Oct. 2003.
42. G. Scutari and S. Barbarossa, "Distributed space-time coding for regenerative relay networks," *IEEE Trans. Wirel. Commun.*, vol. 4, no. 5, pp. 2387–2399, May 2005.
43. Y. Jing and B. Hassibi, "Distributed space-time coding in wireless relay networks," *IEEE Trans. Wirel. Commun.*, vol. 5, no. 12, pp. 3524–3536, Dec. 2006.

44. A. Bletsas, H. Shin, M. Z. Win, and A. Lippman, "A simple cooperative diversity method based on network path selection," *IEEE J. Select. Areas Commun.*, vol. 24, no. 3, pp. 659–672, Mar. 2006.
45. Y. Zou, J. Zhu, B. Zheng, and Y.-D. Yao, "An adaptive cooperation diversity scheme with best-relay selection in cognitive radio networks," *IEEE Trans. Sig. Process.*, vol. 58, no. 10, pp. 5438–5445, Oct. 2010.
46. Y. Zou, Y.-D. Yao, and B. Zheng, "Cognitive transmissions with multiple relays in cognitive radio networks," *IEEE Trans. Wirel. Commun.*, vol. 10, no. 2, pp. 648–659, Feb. 2011.
47. Y. Zou, X. Wang, and W. Shen, "Optimal relay selection for physical-layer security in cooperative wireless networks," *IEEE J. Sel. Areas Commun.*, vol. 31, no. 10, pp. 2099–2111, Oct. 2013.
48. Y. Zou, J. Zhu, X. Li, and L. Hanzo, "Relay selection for wireless communications against eavesdropping: A security-reliability tradeoff perspective," *IEEE Net.*, accepted to appear, Oct. 2015.

Chapter 2
Relay Selection for Enhancing Wireless Secrecy Against Eavesdropping

Abstract In this chapter, we consider a wireless network consisting of a source and a destination with the aid of multiple relay nodes, where an eavesdropper is assumed with an intention to tap the confidential transmission from the source to destination. Considering multiple relays available, we present a relay selection scheme to protect the source-destination transmission against eavesdropping, where only the single "best" relay is selected to help the source transmit the signal to the destination. For comparison purposes, we also consider the conventional direct transmission and random relay selection as benchmark schemes. As the name implies, the direct transmission allows the source to directly transmit its signal to the destination without relying on the relays. By contrast, in the random relay selection, a relay is randomly selected to assist the source-destination transmission. Closed-form intercept probability expressions are derived for the conventional direct transmission and random relay selection as well as the proposed relay selection schemes over Rayleigh fading channels. We also present the secrecy diversity analysis of these there schemes and show that the proposed relay selection obtains the full secrecy diversity, whereas the direct transmission and random relay selection methods achieve the secrecy diversity order of only one. Numerical results demonstrate that the proposed relay selection performs better than both the direct transmission and random relay selection in terms of the intercept probability. Finally, as the number of relays increases, the intercept performance of the proposed relay selection improves significantly, showing the security benefits of exploiting the relay selection to defend against eavesdropping.

2.1 System Model and Problem Formulation

Recently, extensive efforts have been devoted to the research of cooperative relays for improving the performance of wireless networks from different perspectives, including the network coverage [1–3], transmission reliability [4–10] and spectrum utilization [11–15]. More specifically, in [1], the authors studied the deployment of relays for the sake of maximizing the wireless network coverage for a given data transmission rate in Gaussian relay channels, where the network coverage is shown to be sensitive to the locations of relays and the path loss. Later on, in [2] and [3], the network coverage region was developed for an amplify-and-

© Springer International Publishing Switzerland 2016

Y. Zou, J. Zhu, *Physical-Layer Security for Cooperative Relay Networks*,
Wireless Networks, DOI 10.1007/978-3-319-31174-6_2

forward (AF) relay network strategy and the relay location has been optimized for maximizing the coverage region. Alternatively, the employment of cooperative relays was studied for enhancing the reliability of wireless transmissions in [4] and [5], where several relay protocols are proposed, namely the fixed relay, incremental relay and selective relay. It was shown in [6–10] that the outage probability of wireless communications relying on the proposed relay protocols can be largely reduced as compared to conventional direct transmission. Additionally, in [11–15], cooperative relays were exploited in cognitive wireless networks for enabling dynamic spectrum sharing between the primary and secondary users so that a better spectrum utilization can be achieved.

The relay selection technique was first examined to combat the wireless fading effect and thus improve the wireless reliability e.g. [16–22], where only the single "best" relay is selected for participating in forwarding the signal transmission from a source node to its destination. In this way, only two orthogonal channels (namely the source-to-destination channel and the best-relay-to-destination channel) are needed regardless of the number of cooperative relays. Meanwhile, it was shown in [16–22] that the full diversity gain can be achieved by the best relay selection approach, showing its significant advantage in terms of enhancing the wireless reliability. More recently, there are some research works on improving the wireless physical-layer security by using cooperative relays. For example, In [23–25], the authors studied the physical-layer security of wireless communications in the face of an eavesdropper with the help of a relay node, where the AF, decode-and-forward (DF), and compress-and-forward (CF) protocols are investigated and compared with each other. Moreover, a so-called noise-forwarding approach was devised in [26], where a relay node is employed to send an artificially-designed noise for confusing the eavesdropper without affecting the legitimate receiver.

In this chapter, we investigate the physical-layer security of a cooperative wireless network consisting of a source transmitting to its destination with the aid of multiple DF relays, where an eavesdropper is considered for tapping the source-destination transmissions. The main contributions of this chapter can be summarized as follows. First, we propose a single relay selection scheme, where only the "best" relay is chosen among the multiple DF relays for assisting the source-destination transmission against eavesdropping. For comparison purposes, we also consider the conventional direct transmission and random relay selection as our benchmarks. Second, we analyze the closed-form intercept probability performance for the conventional direct transmission, random relay selection and proposed relay selection over Rayleigh fading channels. Finally, some numerical simulations are carried out, showing the advantage of the proposed relay selection over the direct transmission and random relay selection in terms of the intercept probability.

We first present the system model of a wireless relay network in the face of an eavesdropper and then formulate the physical-layer security problem for wireless relay transmissions. As shown in Fig. 2.1, we consider a wireless relay network, where multiple decode-and-forward (DF) relays are available to assist the transmission from a source (S) to its destination (D). Presently, the relay architecture

Fig. 2.1 A source (S) transmitting to its destination (D) with the aid of N relays in the face of an eavesdropper (E)

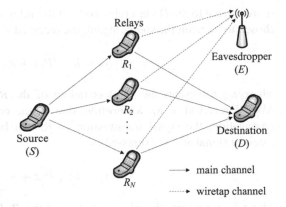

Relays

R_1

Eavesdropper
(E)

R_2

$\vdots$

Destination
(D)

Source
(S)

R_N

→ main channel

⤍ wiretap channel

has been adopted in various commercial wireless network standards e.g. IMT-advanced and IEEE 802.16j. For notational convenience, the set of DF relays is denoted by $\boldsymbol{R} = \{R_1, R_2, \cdots, R_N\}$.

As depicted in Fig. 2.1, when S transmits its message denoted by x to D with the help of N relays, an eavesdropper (E) appears and intends to tap the source transmission. Throughout this book, we assume that both the D and E are beyond the coverage of S and a relay node is exploited to forward the source signal x to D. Although the D becomes capable of receiving the source signal from the relay node, the E can also overhear and decode the relay transmission for the sake of intercepting the source signal. Note that the solid and dash lines represent the main channel and wiretap channel, respectively. Assume that all the wireless links between any two network nodes of Fig. 2.1 are modeled as independent Rayleigh fading channels. Moreover, all the receivers are considered to have the zero-mean additive white Gaussian noise (AWGN) with a variance of N_0.

Following the physical-layer security literature, the E is assumed to know everything about the S-D transmission, including the waveform, carrier frequency and bandwidth, encoding and modulation, except that the source message x is confidential. We assume that S transmits its signal x at a power of P_s. Thus, the signal received at a relay node denoted by R_i is expressed as

$$y_i = h_{si}\sqrt{P_s}x + n_i, \tag{2.1}$$

where h_{si} represents the wireless fading of the S-R_i channel and n_i represents AWGN received at R_i. Using the Shannon's capacity formula, the capacity of the S-R_i channel is obtained from (2.1) as

$$C_{si} = \frac{1}{2}\log_2(1 + |h_{si}|^2\gamma_s), \tag{2.2}$$

where $\gamma_s = P_s/N_0$ represents the transmit signal-to-noise ratio (SNR) and the factor $\frac{1}{2}$ is due to the fact that two orthogonal time slots are required for transmitting the source signal to the D via a relay node. If the relay R_i succeeds in decoding x and is chosen to transmit the source signal, the received signal at D can be given by

$$y_d = h_{id} \sqrt{P_s} x + n_d, \tag{2.3}$$

where h_{id} represents the wireless fading of the R_i-D channel and n_d represents AWGN received at R_i. Meanwhile, due to the openness nature of the wireless propagation, the signal transmission of R_i can be overheard by E. Hence, the received signal at E is expressed as

$$y_e = h_{ie} \sqrt{P_s} x + n_e, \tag{2.4}$$

where h_{ie} represents the wireless fading of the R_i-E channel and n_e represents the AWGN at E. From (2.3), the capacity of the R_i-D channel is obtained as

$$C_{id} = \frac{1}{2} \log_2(1 + |h_{id}|^2 \gamma_s). \tag{2.5}$$

Similarly, from (2.3), the capacity of the R_i-E channel is given by

$$C_{ie} = \frac{1}{2} \log_2(1 + |h_{ie}|^2 \gamma_s). \tag{2.6}$$

As discussed in [26], when the channel capacity C_{id} is higher than C_{ie}, a perfect secrecy can be achieved in an information-theoretic sense. However, if C_{id} is lower than C_{ie}, the information-theoretic security becomes unachievable. In this case, the E would succeed in intercepting the transmission of R_i and an intercept event is viewed to occur. Additionally, the probability of the E intercepting the source message is referred to as the *intercept probability*, which is used to measure the performance of wireless physical-layer security. We focus on exploring the relay selection for improving the wireless security in terms of decreasing the intercept probability.

2.2   Relay Selection for Secrecy Enhancement

In this section, we propose a relay selection scheme for enhancing the wireless secrecy against eavesdropping and analyze the intercept probability of the proposed scheme over Rayleigh fading channels. For comparison purposes, we also present the conventional direct transmission and random relay selection methods.

2.2.1 Direct Transmission

As a baseline, let us consider the conventional direct transmission from S to D without relying on relays. Considering that S transmits its signal directly to D at a power of P_s, we can obtain the capacity of the S-D channel as

$$C_{sd} = \log_2(1 + |h_{sd}|^2 \gamma_s), \tag{2.7}$$

where h_{sd} represents the wireless fading of the S-D channel. Meanwhile, the broadcast nature of wireless transmission leads to the fact that the E may overhear the source transmission. Similarly, the capacity of the S-E channel is given by

$$C_{se} = \log_2(1 + |h_{se}|^2 \gamma_s), \tag{2.8}$$

where h_{se} represents the wireless fading of the S-E channel. As aforementioned, an intercept event happens when the capacity of the main channel C_{sd} falls below that of the wiretap channel C_{se}. Therefore, using (2.7) and (2.8), the intercept probability experienced at the E relying on the direct transmission scheme is obtained as

$$P_{\text{int}}^{\text{direct}} = \Pr(C_{sd} < C_{se}) = \Pr\left(|h_{sd}|^2 < |h_{se}|^2\right). \tag{2.9}$$

It needs to be pointed out that the channel coefficients h_{sd} and h_{se} are modeled as independent Rayleigh fading, implying that the magnitudes $|h_{sd}|^2$ and $|h_{se}|^2$ are independent exponential random variables. Letting σ_{sd}^2 and σ_{se}^2 denote the means of $|h_{sd}|^2$ and $|h_{se}|^2$, we obtain the intercept probability of the conventional direct transmission scheme as

$$P_{\text{int}}^{\text{direct}} = \frac{\sigma_{se}^2}{\sigma_{sd}^2 + \sigma_{se}^2}. \tag{2.10}$$

2.2.2 Random Relay Selection

In this subsection, we present the random relay selection as a benchmark, where a relay node is randomly selected among the set R to assist the S-D transmission. Without loss of generality, we consider that the relay node R_i is selected for forwarding the source transmission. To be specific, the S first transmits its signal x to the relay R_i, which then decodes its received signal. If the relay R_i succeeds in decoding x, it forwards x to the D. Otherwise, the relay node keeps silent.

For notational convenience, let $\phi = 0$ denote the case that the relay R_i succeeds in decoding and $\phi = 1$ denote the case that the relay fails. According to the Shannon's coding theorem, if the channel capacity C_{si} is higher than the transmission rate of

x, the relay R_i is able to successfully decode the source signal x. However, if C_{si} falls below the transmission rate, it is impossible for the relay to recover the source signal. Therefore, the cases $\phi = 0$ and $\phi = 1$ can be described as

$$\phi = 0 : \ C_{si} < R \tag{2.11}$$

$$\phi = 1 : \ C_{si} > R, \tag{2.12}$$

where R denotes the transmission rate of x and C_{si} is given by (2.2). If the relay R_i succeeds in decoding, it then forwards the source signal x to the D, which may be intercepted by the E. As discussed in [26], if the capacity of the R_i-D channel is lower than that of the R_i-E, the information-theoretic security cannot be achieved and an intercept event occurs in this case. Thus, given that R_i is selected and succeeds in decoding, the intercept probability experienced at the E is expressed as

$$P_{\text{int}}(R_i, \varphi = 1) = \Pr(C_{id} < C_{ie}, C_{si} > R), \tag{2.13}$$

where C_{id} and C_{ie} are given by (2.5) and (2.6), respectively. In the random relay selection scheme, each relay node is randomly selected in an equal manner. Therefore, using the law of total probability, the intercept probability of random relay selection is given by

$$P_{\text{int}}^{\text{random}} = \frac{1}{N} \sum_{i=1}^{N} P_{\text{int}}(R_i, \varphi = 1) = \frac{1}{N} \sum_{i=1}^{N} \Pr(C_{id} < C_{ie}, C_{si} > R), \tag{2.14}$$

where N is the number of relays. Substituting C_{si}, C_{id} and C_{ie} from (2.2), (2.5) and (2.6) into (2.14) yields

$$P_{\text{int}}^{\text{random}} = \frac{1}{N} \sum_{i=1}^{N} \Pr(|h_{id}|^2 < |h_{ie}|^2, |h_{si}|^2 > \Delta), \tag{2.15}$$

where $\Delta = \frac{2^{2R}-1}{\gamma_s}$. Noting that $|h_{si}|^2$, $|h_{id}|^2$ and $|h_{ie}|^2$ are independent exponentially distributed random variables with respective means of σ_{si}^2, σ_{id}^2 and σ_{ie}^2, we obtain a close-form expression of the intercept probability of random relay selection as

$$P_{\text{int}}^{\text{random}} = \frac{1}{N} \sum_{i=1}^{N} \frac{\sigma_{ie}^2}{\sigma_{id}^2 + \sigma_{ie}^2} \exp(-\frac{\Delta}{\sigma_{si}^2}), \tag{2.16}$$

where $\exp(\cdot)$ denotes the exponential function.

2.2.3 Proposed Relay Selection

This subsection proposes a relay selection scheme for enhancing the security of S-D transmissions against eavesdropping. To be specific, S first transmits its signal x to the N relays, which then decode their received signals for recovering the source signal x. Here, the set of relays that succeeds in recovering the source signal is called the decoding set and denoted by $\mathscr{D}$. Given N relays, we have 2^N possible combinations for the decoding set $\mathscr{D}$. Hence, the sample space of the decoding set is expressed as

$$\Omega = \{\emptyset, \mathscr{D}_1, \mathscr{D}_2, \cdots, \mathscr{D}_n, \cdots, \mathscr{D}_{2^N-1}\}, \tag{2.17}$$

where $\emptyset$ represents a null set and $\mathscr{D}_n$ represents the n-th non-empty subcollection of the N relays. According to the Shannon's coding theorem, when the channel capacity falls below the transmission rate, a receiver would fail to decode the source signal [27]. Hence, the event $\mathscr{D} = \emptyset$ means that all relay nodes fail to decode the source signal, which can be described as

$$C_{si} < R, \quad i = 1, 2, \cdots, N, \tag{2.18}$$

where C_{si} is given by (2.2) and R is the transmission rate of the source signal. Meanwhile, the event $\mathscr{D} = \mathscr{D}_n$ implies that the relays within $\mathscr{D}_n$ succeed in decoding and the others within $\bar{\mathscr{D}}_n$ fail to decode, where $\bar{\mathscr{D}}_n$ represents the complementary set of $\mathscr{D}_n$. Similarly, in an information-theoretic sense, the event $\mathscr{D} = \mathscr{D}_n$ is described as

$$C_{si} > R, \quad i \in \mathscr{D}_n \tag{2.19}$$

$$C_{sj} < R, \quad j \in \bar{\mathscr{D}}_n. \tag{2.20}$$

If the decoding set $\mathscr{D}$ is null, meaning that all relays fail to decode x, no relay is chosen to forward the source transmission and both the D and E are unable to decode the source message in this case. If the decoding set is a non-empty set $\mathscr{D}_n$, a specific relay is chosen from $\mathscr{D}_n$ for forwarding its decoded signal x to D. Since the E is passive, it is challenging to know its channel state information (CSI) in practice. As a consequence, we assume that only the CSI of the main channel spanning from the relay nodes to D is available without knowing the E's CSI. Under this condition, a relay that succeeds in decoding x and has the highest capacity of the main channel C_{id} is typically considered as the "best" node to forward the source signal x to the D. Hence, from (2.5), the relay selection criterion can be written as

$$\text{Best Relay} = \arg\max_{i \in \mathscr{D}_n} C_{id} = \arg\max_{i \in \mathscr{D}_n} |h_{id}|^2, \tag{2.21}$$

which shows that only the $|h_{id}|^2$ is needed in carrying out the relay selection without requiring the E's CSI $|h_{ie}|^2$. Using the law of total probability, we can obtain the intercept probability of the proposed relay selection as

$$P_{\text{int}}^{\text{proposed}} = \sum_{n=1}^{2^N-1} \Pr(C_{bd} < C_{be}, \mathcal{D} = \mathcal{D}_n), \qquad (2.22)$$

where C_{bd} and C_{be} represent the channel capacity from the "best" relay to the D and E, respectively. Noting that each node within the decoding set $\mathcal{D}_n$ has a possibility of becoming the "best" relay, we can rewrite (2.22) as

$$P_{\text{int}}^{\text{proposed}} = \sum_{n=1}^{2^N-1} \Pr(\mathcal{D} = \mathcal{D}_n) \sum_{i \in \mathcal{D}_n} \Pr(C_{id} < C_{ie}, b = i), \qquad (2.23)$$

where b denotes the "best" relay node. From (2.19) and (2.20), the term $\Pr(\mathcal{D} = \mathcal{D}_n)$ is given by

$$\Pr(\mathcal{D} = \mathcal{D}_n) = \prod_{i \in \mathcal{D}_n} \Pr(C_{si} > R) \prod_{j \in \bar{\mathcal{D}}_n} \Pr(C_{sj} < R). \qquad (2.24)$$

Combining (2.2) and (2.24) gives

$$\Pr(\mathcal{D} = \mathcal{D}_n) = \prod_{i \in \mathcal{D}_n} \Pr(|h_{si}|^2 > \Delta) \prod_{j \in \bar{\mathcal{D}}_n} \Pr(|h_{sj}|^2 < \Delta). \qquad (2.25)$$

Noting that $|h_{si}|^2$ is an exponentially distributed random variable, we havev

$$\Pr(\mathcal{D} = \mathcal{D}_n) = \prod_{i \in \mathcal{D}_n} \exp(-\frac{\Delta}{\sigma_{si}^2}) \prod_{j \in \bar{\mathcal{D}}_n} [1 - \exp(-\frac{\Delta}{\sigma_{sj}^2})], \qquad (2.26)$$

where σ_{si}^2 and σ_{sj}^2 are the means of $|h_{si}|^2$ and $|h_{sj}|^2$, respectively. Using (2.21), we can obtain

$$\Pr(C_{id} < C_{ie}, b = i) = \Pr(C_{id} < C_{ie}, |h_{id}|^2 > \max_{j \in \{\mathcal{D}_n - R_i\}} |h_{jd}|^2), \qquad (2.27)$$

where '−' denotes the set difference. Substituting C_{id} and C_{ie} from (2.5) and (2.6) into (2.27) yields

$$\Pr(C_{id} < C_{ie}, b = i) = \Pr(|h_{ie}|^2 > |h_{id}|^2, \max_{j \in \{\mathcal{D}_n - R_i\}} |h_{jd}|^2 < |h_{id}|^2). \qquad (2.28)$$

Noting again that $|h_{ie}|^2$, $|h_{id}|^2$ and $|h_{jd}|^2$ are independent exponential random variables with respective means of σ_{ie}^2, σ_{id}^2 and σ_{jd}^2, we can compute (2.28) as

$$\Pr(C_{id} < C_{ie}, b = i) = \int_0^\infty \frac{1}{\sigma_{id}^2} \exp(-\frac{x}{\sigma_{id}^2} - \frac{x}{\sigma_{ie}^2}) \prod_{j \in \{\mathcal{D}_n - R_i\}} [1 - \exp(-\frac{x}{\sigma_{jd}^2})] dx, \qquad (2.29)$$

where the item $\prod_{j\in\{\mathscr{D}_n-R_i\}} [1 - \exp(-\frac{x}{\sigma_{jd}^2})]$ can be expanded by using the binomial theorem as

$$\prod_{j\in\{\mathscr{D}_n-R_i\}} [1 - \exp(-\frac{x}{\sigma_{jd}^2})] = 1 + \sum_{m=1}^{2^{|\mathscr{D}_n|-1}-1} (-1)^{|\mathscr{C}_{n,m}|} \exp(-\sum_{j\in\mathscr{C}_{n,m}} \frac{x}{\sigma_{jd}^2}), \qquad (2.30)$$

where $\mathscr{C}_{n,m}$ denotes the m-th non-empty collection of the set $\mathscr{D}_n$ and $|\cdot|$ denotes the set cardinality. Combining (2.29) and (2.30) gives

$$\Pr(C_{id} < C_{ie}, b = i) = \int_0^\infty \frac{1}{\sigma_{id}^2} \exp(-\frac{x}{\sigma_{id}^2} - \frac{x}{\sigma_{ie}^2})dx$$

$$+ \sum_{m=1}^{2^{|\mathscr{D}_n|-1}-1} \int_0^\infty \frac{(-1)^{|\mathscr{C}_{n,m}|}}{\sigma_{id}^2} \exp(-\frac{x}{\sigma_{id}^2} - \frac{x}{\sigma_{ie}^2} - \sum_{j\in\mathscr{C}_{n,m}} \frac{x}{\sigma_{jd}^2})dx \qquad (2.31)$$

which can be further obtained as

$$\Pr(C_{id} < C_{ie}, b = i) = \frac{\sigma_{ie}^2}{\sigma_{ie}^2 + \sigma_{id}^2}$$

$$+ \sum_{m=1}^{2^{|\mathscr{D}_n|-1}-1} (-1)^{|\mathscr{C}_{n,m}|} (1 + \frac{\sigma_{id}^2}{\sigma_{ie}^2} + \sum_{j\in\mathscr{C}_{n,m}} \frac{\sigma_{id}^2}{\sigma_{jd}^2})^{-1}. \qquad (2.32)$$

Finally, substituting $\Pr(\mathscr{D} = \mathscr{D}_n)$ and $\Pr(C_{id} < C_{ie}, b = i)$ from (2.26) and (2.32) into (2.23) yields a closed-form expression of the intercept probability of proposed relay selection scheme.

2.3 Asymptotic Intercept Probability and Secrecy Diversity

In this section, we conduct the asymptotic intercept probability analysis of the conventional direct transmission and random relay selection as well as the proposed relay selection schemes. Although the closed-form expression of (2.23) can be used to numerically evaluate the intercept probability of proposed relay selection, it fails to provide an insight into the impact of the number of relays on the security performance. To this end, we now focus on carrying out an asymptotic intercept probability analysis for characterizing the wireless security performance.

Let us first consider the conventional direct transmission scheme. As observed from (2.10), the intercept probability of direct transmission only relates to the average channel gains σ_{sd}^2 and σ_{se}^2. Denoting $\lambda_{me} = \sigma_{sd}^2/\sigma_{se}^2$, we can rewrite (2.10) as

$$P_{\text{int}}^{\text{direct}} = \frac{1}{1 + \lambda_{me}^{-1}} \cdot \frac{1}{\lambda_{me}}, \qquad (2.33)$$

where λ_{me} is the ratio of an average gain of the main channel (spanning from S to D) to that of the eavesdropper's wiretap channel (spanning from S to E), called the main-to-eavesdropper ratio (MER) [28]. It can be observed from (2.33) that as the MER λ_{me} increases to infinity, the intercept probability behaves as $\frac{1}{\lambda_{me}}$, showing a decrease of the intercept probability with an increasing MER. As discussed in [28–30], the secrecy diversity is defined as an asymptotic ratio of the intercept probability to MER as $\lambda_{me} \to \infty$ i.e.

$$d = - \lim_{\lambda_{me} \to \infty} \frac{\log(P_{\text{int}})}{\log(\lambda_{me})}. \tag{2.34}$$

Kindly note that a high MER can be achieved by exploiting some signal processing techniques (e.g. beamforming) so that a constructive interference is encountered at the main channel, whereas the wiretap channel experiences a destructive interference. Combining (2.33) and (2.34), we obtain the secrecy diversity of conventional direct transmission scheme as

$$d_{\text{direct}} = - \lim_{\lambda_{me} \to \infty} \frac{\log(P_{\text{int}}^{\text{direct}})}{\log(\lambda_{me})} = 1, \tag{2.35}$$

which shows that only the diversity order of one is achieved by the direct transmission. Similarly, denoting $\sigma_{id}^2 = \alpha_{id}\sigma_{sd}^2$ and $\sigma_{ie}^2 = \alpha_{ie}\sigma_{se}^2$, we rewrite (2.16) as

$$P_{\text{int}}^{\text{random}} = \frac{1}{N} \sum_{i=1}^{N} \frac{\alpha_{ie}}{\alpha_{id} + \alpha_{ie}\lambda_{me}^{-1}} \exp(-\frac{\Delta}{\sigma_{si}^2}) \cdot \frac{1}{\lambda_{me}}, \tag{2.36}$$

which also shows that as the MER approaches to infinity, the intercept probability of random relay selection behaves as $\frac{1}{\lambda_{me}}$. Combining (2.34) and (2.36), we obtain the secrecy diversity of random relay selection as

$$d_{\text{random}} = - \lim_{\lambda_{me} \to \infty} \frac{\log(P_{\text{int}}^{\text{random}})}{\log(\lambda_{me})} = 1, \tag{2.37}$$

from which the random relay selection scheme achieves the same secrecy diversity as the conventional direct transmission, showing that no secrecy benefits are achieved by the random relay selection. In what follows, we characterize the secrecy diversity order of the proposed relay selection through an asymptotic intercept probability analysis in high MER region. Let us consider $Y_j = 1 - \exp(-\frac{x}{\sigma_{jd}^2})$, where x is a random variable having the probability density function (PDF) as

$$p(x) = \frac{1}{\sigma_{id}^2} \exp(-\frac{x}{\sigma_{id}^2} - \frac{x}{\sigma_{ie}^2}), \tag{2.38}$$

for $x \geq 0$. Using (2.38), we can obtain the mean of Y_j as

$$E(Y_j) = \int_0^{\infty} \frac{1}{\sigma_{id}^2} \exp(-\frac{x}{\sigma_{id}^2} - \frac{x}{\sigma_{ie}^2}) dx - \int_0^{\infty} \frac{1}{\sigma_{id}^2} \exp(-\frac{x}{\sigma_{id}^2} - \frac{x}{\sigma_{ie}^2} - \frac{x}{\sigma_{jd}^2}) dx$$

$$= \frac{\sigma_{ie}^2}{\sigma_{id}^2 + \sigma_{ie}^2} - \frac{\sigma_{jd}^2 \sigma_{ie}^2}{\sigma_{id}^2 \sigma_{jd}^2 + \sigma_{id}^2 \sigma_{ie}^2 + \sigma_{jd}^2 \sigma_{ie}^2}. \quad (2.39)$$

By denoting $\sigma_{id}^2 = \alpha_{id}\sigma_{sd}^2$, $\sigma_{jd}^2 = \alpha_{jd}\sigma_{sd}^2$, $\sigma_{ie}^2 = \alpha_{ie}\sigma_{se}^2$, (2.39) is rewritten as

$$E(Y_j) = \left(\frac{\alpha_{ie}}{\alpha_{id} + \alpha_{ie}\lambda_{me}^{-1}} - \frac{\alpha_{jd}\alpha_{ie}}{\alpha_{id}\alpha_{jd} + \alpha_{id}\alpha_{ie}\lambda_{me}^{-1} + \alpha_{jd}\alpha_{ie}\lambda_{me}^{-1}}\right) \cdot \frac{1}{\lambda_{me}}, \quad (2.40)$$

where $\lambda_{me} = \sigma_{sd}^2/\sigma_{se}^2$. Meanwhile, the mean of Y_j^2 is computed as

$$E(Y_j^2) = \left(\frac{\alpha_{ie}}{\alpha_{id} + \alpha_{ie}\lambda_{me}^{-1}} - \frac{2\alpha_{jd}\alpha_{ie}}{\alpha_{id}\alpha_{jd} + \alpha_{ie}\alpha_{id}\lambda_{me}^{-1} + \alpha_{jd}\alpha_{ie}\lambda_{me}^{-1}}\right) \cdot \frac{1}{\lambda_{me}}$$

$$+ \frac{\alpha_{jd}\alpha_{ie}}{\alpha_{id}\alpha_{jd} + 2\alpha_{ie}\alpha_{id}\lambda_{me}^{-1} + \alpha_{jd}\alpha_{ie}\lambda_{me}^{-1}} \cdot \frac{1}{\lambda_{me}}. \quad (2.41)$$

As shown in (2.40) and (2.41), as the MER λ_{me} tends infinity, both $E(Y_j)$ and $E(Y_j^2)$ approaches to zero, implying that the random variable Y_j goes to zero with probability one for $\lambda_{me} \to +\infty$. Therefore, letting $\lambda_{me} \to +\infty$ and using the Taylor series expansion, we can obtain

$$Y_j = 1 - \exp(-\frac{x}{\sigma_{jd}^2}) = \frac{x}{\sigma_{jd}^2} + O(\frac{1}{\lambda_{me}}), \quad (2.42)$$

where $O(\cdot)$ represents a higher-order infinitesimal. Combining (2.29) and (2.42) and ignoring the infinitesimal, we have

$$\Pr(C_{id} < C_{ie}, b = i) = \frac{1}{\sigma_{id}^2} \prod_{j \in \{\mathscr{D}_n - R_i\}} \frac{1}{\sigma_{jd}^2} \int_0^{\infty} x^{|\mathscr{D}_n| - 1} \exp(-\frac{x}{\sigma_{id}^2} - \frac{x}{\sigma_{ie}^2}) dx, \quad (2.43)$$

for $\lambda \to +\infty$. Performing the integral of (2.43) yields

$$\Pr(C_{id} < C_{ie}, b = i) = \frac{(|\mathscr{D}_n| - 1)!}{\sigma_{id}^2} \prod_{j \in \{\mathscr{D}_n - R_i\}} \frac{1}{\sigma_{jd}^2} (\frac{1}{\sigma_{id}^2} + \frac{1}{\sigma_{ie}^2})^{-|\mathscr{D}_n|}, \quad (2.44)$$

which can be rewritten as

$$\Pr(C_{id} < C_{ie}, b = i) = \frac{(|\mathscr{D}_n|-1)!}{\alpha_{id}}\left(\frac{\alpha_{id}\alpha_{ie}}{\alpha_{id}+\alpha_{ie}\lambda_{me}^{-1}}\right)^{|\mathscr{D}_n|}$$

$$\times \prod_{j\in\{\mathscr{D}_n-R_i\}} \frac{1}{\alpha_{jd}} \cdot \left(\frac{1}{\lambda_{me}}\right)^{|\mathscr{D}_n|}, \qquad (2.45)$$

for $\lambda \to +\infty$. Substituting $\Pr(C_{id} < C_{ie}, b = i)$ from (2.45) into (2.23) gives

$$P_{\text{int}}^{\text{proposed}} = \sum_{n=1}^{2^N-1} \Pr(D = \mathscr{D}_n) \sum_{i\in\mathscr{D}_n} \frac{(|\mathscr{D}_n|-1)!}{\alpha_{id}}\left(\frac{\alpha_{id}\alpha_{ie}}{\alpha_{id}+\alpha_{ie}\lambda_{me}^{-1}}\right)^{|\mathscr{D}_n|}$$

$$\prod_{j\in\{\mathscr{D}_n-R_i\}} \frac{1}{\alpha_{jd}} \cdot \left(\frac{1}{\lambda_{me}}\right)^{|\mathscr{D}_n|}, \qquad (2.46)$$

where $\Pr(D = \mathscr{D}_n)$ is given by (2.26). It is shown from (2.46) that the secrecy diversity order of proposed relay selection relates to the SNR γ_s as well as the average gains of the main channel and wiretap channel. Considering $\lambda_s \to \infty$, we can further simplify (2.46) as

$$P_{\text{int}}^{\text{proposed}} = \sum_{i=1}^{N} \frac{(N-1)!}{\alpha_{id}}\left(\frac{\alpha_{id}\alpha_{ie}}{\alpha_{id}+\alpha_{ie}\lambda_{me}^{-1}}\right)^{N} \prod_{j=1, j\neq i}^{N} \frac{1}{\alpha_{jd}} \cdot \left(\frac{1}{\lambda_{me}}\right)^{N}, \qquad (2.47)$$

where N is the number of relays. Combining (2.34) and (2.47), we obtain the secrecy diversity order of proposed relay selection as

$$d_{\text{proposed}} = -\lim_{\lambda_{me}\to\infty} \frac{\log(P_{\text{int}}^{\text{proposed}})}{\log(\lambda_{me})} = N, \qquad (2.48)$$

for $\lambda_s \to \infty$. As shown in (2.48), the secrecy diversity order of proposed relay selection is equal to the number of relays, showing a significant secrecy benefits achieved by the proposed scheme, as compared to the conventional direct transmission and random relay selection.

2.4 Numerical Results and Discussions

This section presents numerical intercept probability results of the conventional direct transmission, random relay selection and proposed relay selection schemes. To be specific, the intercept probabilities of these schemes are obtained by plotting (2.10), (2.16) and (2.23). In our numerical results, the average gains of the main channel (spanning from S via relays to D) are given by $\sigma_{sd}^2 = \sigma_{si}^2 = \sigma_{id}^2 = 1$ and the average gains of the wiretap channel (spanning from relays to E) are specified

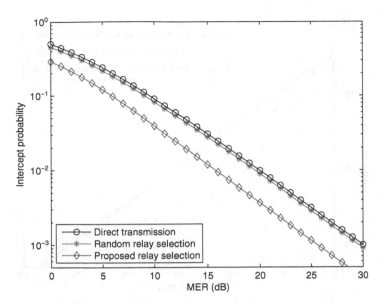

Fig. 2.2 Intercept probability versus MER λ_{me} of the conventional direct transmission and random relay selection as well as the proposed relay selection schemes

to $\sigma_{ie}^2 = 0.1$, unless otherwise mentioned. In addition, an SNR of $\gamma_s = 10\,\text{dB}$, a transmission rate of $R = 1\,\text{bit/s/Hz}$, and $N = 4$ are considered, unless otherwise stated.

Figure 2.2 shows the intercept probability comparison among the conventional direct transmission, random relay selection and proposed relay selection schemes by plotting (2.10), (2.16) and (2.23) as a function of λ_{me}. It is shown from Fig. 2.2 that with an increasing MER, the intercept probabilities of the direct transmission, random relay selection and proposed relay selection schemes are reduced significantly. This is due to the fact that as the MER λ_{me} increases, the legitimate D has a better signal reception from the S, whereas the quality of signal reception at the E becomes worse, which leads to an intercept probability improvement. Figure 2.2 also shows that the conventional direct transmission is the worst and the proposed relay selection performs the best in terms of the intercept probability. This confirms the secrecy benefits achieved by the proposed relay selection, as compared to the conventional direct transmission and random relay selection.

Figure 2.3 depicts the intercept probability versus MER λ_{me} of the proposed relay selection scheme for different SNRs γ_s. It can be seen from Fig. 2.3 that as the SNR increases from $\gamma_s = 5$ to $12\,\text{dB}$, the intercept probability of the proposed relay selection scheme decreases accordingly. By contrast, one can observe from (2.10) that the intercept probability of the conventional direct transmission only relates to the average channel gains of σ_{sd}^2 and σ_{se}^2 and has nothing to do with the SNR γ_s. This means that with an increasing SNR, the intercept probability of

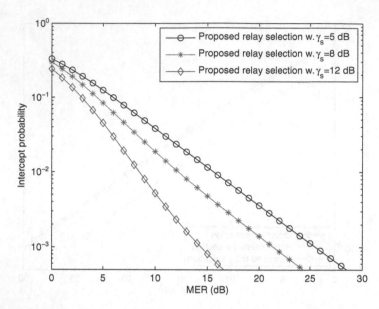

Fig. 2.3 Intercept probability versus MER λ_{me} of the proposed relay selection scheme for different SNRs γ_s

the direct transmission remains unchanged. Therefore, as the SNR increases, the intercept probability improvement of the proposed relay selection scheme over the conventional direct transmission becomes more significant.

In Fig. 2.4, we demonstrate the intercept probability versus λ_{me} of the random relay selection and proposed relay selection for different number of relays N. As shown in Fig. 2.4, upon increasing the number of relays from $N = 2$ to 6, the intercept probability of random relay selection scheme keeps unchanged, showing no secrecy benefits achieved by the random relay selection. By contrast, the intercept performance of proposed relay selection significantly improves when the number of relays increases from $N = 2$ to 6. This shows that the physical-layer security of wireless transmissions relying on the proposed relay selection scheme can be significantly enhanced by exploiting more relays. Additionally, one can also observe from Fig. 2.4 that for both the cases of $N = 2$ and $N = 6$, the proposed relay selection strictly outperforms the random relay selection in terms of the intercept probability.

Figure 2.5 shows the comparison between the exact intercept probability and asymptotic intercept probability for the proposed relay selection scheme by plotting (2.23) and (2.47) as a function of the MER λ_{me}. One can observe from Fig. 2.5 that for all the cases of $N = 2$, 4 and 6, the asymptotic intercept probability curves converge to the corresponding exact intercept probability results in high MER region, showing the correctness of the asymptotic intercept analysis. Moreover, as the number of relays increases from $N = 2$ to 6, the slopes of the intercept

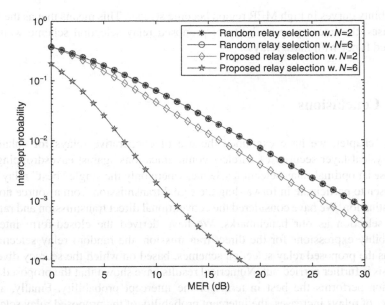

Fig. 2.4 Intercept probability versus MER λ_{me} of the conventional random relay selection and the proposed relay selection schemes for different number of relays

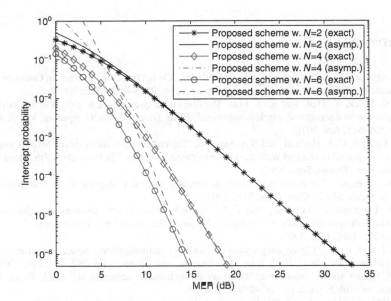

Fig. 2.5 Exact and asymptotic intercept probability comparison for the proposed relay selection scheme for different number of relays with $\gamma_s = 50$ dB

probability curves in high MER region become steeper. This means that as the MER increases, the intercept probability of proposed relay selection scheme would be reduced faster with more relays.

2.5 Conclusions

In this chapter, we have examined the use of cooperative relays for enhancing the physical-layer security of wireless communications against eavesdropping and propose an optimal relay selection scheme, where only the single "best" relay node is chosen to participate in forwarding the signal transmission from a source node to its destination. We have considered the conventional direct transmission and random relay selection as our benchmarks. We have derived the closed-form intercept probability expressions for the direct transmission, the random relay selection as well as the proposed relay selection schemes, based on which the secrecy diversity analysis is further carried out. Numerical results have shown that the proposed relay selection performs the best in terms of the intercept probability. Finally, as the number of relays increases, the intercept probability of the proposed relay selection scheme improves significantly, showing the security advantage of exploiting relay selection to defend against eavesdropping.

References

1. V. Aggarwal, A. Bennatan, and A.R. Calderbank, "On maximizing coverage in Gaussian relay channels," *IEEE Trans. Inf. Theory*, vol. 55, no. 6, pp. 2518–2536, Jun. 2009.
2. A.K. Sadek, Z. Han, and K. J. Liu, "Distributed relay-assignment protocols for coverage expansion in cooperative wireless networks," *IEEE Trans. Mobile Computing*, vol. 9, no. 4, pp. 505–515, Apr. 2010.
3. B. Razeghi, G.A. Hodtani, and S.A. Seyedin, "Coverage region analysis for MIMO amplify-and-forward relay channel with the source to destination link," in *Proc. 2014 7th Intern. Sym. Telecomm.*, Tehran, Sept. 2014.
4. J.N. Laneman, "Cooperative diversity in wireless networks: Algorithms and architectures, Ph.D. thesis, M.I.T., Cambridge, MA, 2002.
5. J.N. Laneman, D.N.C. Tse, and G.W. Wornell, "Cooperative diversity in wireless networks: Efficient protocols and outage behavior, *IEEE Trans. Inf. Theory*, vol. 50, no. 12, pp. 3062–3080, Dec. 2004.
6. M. Janani, *et al.*, "Coded cooperation in wireless communications: Space-time transmission and iterative decoding," *IEEE Trans. Sig. Process.*, vol. 52, no. 2, pp. 362–371, Feb. 2004.
7. T.E. Hunter and A. Nosratinia, "Diversity through coded cooperation," *IEEE Trans. Wirel. Commun.*, vol. 5, no. 2, pp. 283–289, Feb. 2006.
8. S. Ikki and M.H. Ahmed, "Performance analysis of cooperative diversity wireless networks over Nakagami-m fading channel," *IEEE Commun. Lett.*, vol. 11, no. 4, pp. 334–336, Apr. 2007.
9. Y. Zou, B. Zheng, and W.-P. Zhu, "An opportunistic cooperation scheme and its BER analysis," *IEEE Trans. Wirel. Commun.*, vol. 8, no. 9, pp. 4492–4497, Sept. 2009.

10. Y. Zou, B. Zheng, and J. Zhu, "Outage analysis of opportunistic cooperation over Rayleigh fading channels," *IEEE Trans. Wirel. Commun.*, vol. 8, no. 6, pp. 3077–3085, Jun. 2009.
11. G. Ganesan and Y. Li, "Cooperative spectrum sensing in cognitive radio, Part I: Two user networks," *IEEE Trans. Wirel. Commun.*, vol. 6, no. 6, pp. 2204–2213, Jun. 2007.
12. J. Ma, G. Zhao, and Y. Li, "Soft combination and detection for cooperative spectrum sensing in cognitive radio networks," *IEEE Trans. Wirel. Commun.*, vol. 7, no. 11, pp. 4502–4507, Nov. 2008.
13. Y. Zou, Y.-D. Yao, and B. Zheng, "A selective-relay based cooperative spectrum sensing scheme without dedicated reporting channels in cognitive radio networks," *IEEE Trans. Wirel. Commun.*, vol. 10, no. 4, pp. 1188–1198, Apr. 2011.
14. Y. Zou, Y.-D. Yao, and B. Zheng, "Cognitive transmissions with multiple relays in cognitive radio networks," *IEEE Trans. Wirel. Commun.*, vol. 10, no. 2, pp. 648–659, Feb. 2011.
15. Y. Zou, Y.-D. Yao, and B. Zheng, "Cooperative relay techniques for cognitive radio systems: Spectrum sensing and secondary user transmissions," *IEEE Commun. Mag.*, vol. 50, no. 4, pp. 98–103, Apr. 2012.
16. A. Bletsas, H. Shin, M. Z. Win, and A. Lippman, "A simple cooperative diversity method based on network path selection," *IEEE J. Select. Areas in Commun.*, vol. 24, no. 3, pp. 659–672, Mar. 2006.
17. E. Beres and R. S. Adve, "Selection cooperation in multi-source cooperative networks," *IEEE Trans. Wireless Commun.*, vol. 7, no. 1, pp. 118–127, Jan. 2008.
18. A.S. Ibrahim, *et al.*, "Cooperative communications with relay-selection: when to cooperate and whom to cooperate with," *IEEE Trans. Wirel. Commun.*, vol. 7, no. 7, pp. 2814–2827, Jul. 2008.
19. D.S. Michalopoulos and G.K. Karagiannidis, "Performance analysis of single relay selection in Rayleigh fading," IEEE Trans. Wirel. Commun., vol. 7, no. 10, pp. 3718–3724, Oct. 2008.
20. Y. Jing and H. Jafarkhani, "Single and multiple relay selection schemes and their achievable diversity orders," *IEEE Trans. Wirel. Commun.*, vol. 8, no. 3, pp. 1414–1423, Mar. 2009.
21. Y. Zou, J. Zhu, B. Zheng, and Y.-D. Yao, "An adaptive cooperation diversity scheme with best-relay selection in cognitive radio networks," *IEEE Trans. Sig. Process.*, vol. 58, no. 10, pp. 5438–5445, Oct. 2010.
22. S. Ikki and M. Ahmed, "Performance analysis of adaptive decode-and-forward cooperative diversity networks with best-relay selection," *IEEE Trans. Commun.*, vol. 58, no. 1, pp. 68–72, Oct. 2010.
23. M. Yuksel and E. Erkip, "Secure communication with a relay helping the wiretapper," in *Proc. 2007 IEEE Inf. Theory Workshop*, Lake Tahoe, CA, Sept. 2007.
24. J. Mo, M. Tao, and Y. Liu, "Relay placement for physical layer security: A secure connection perspective," *IEEE Commun. Lett.*, vol. 16, no. 6, pp. 878–881, Jun. 2012.
25. H. Sakran, *et al.*, "Proposed relay selection scheme for physical layer security in cognitive radio networks," *IET Commun.*, vol. 6, no. 16, pp. 2676–2687, Aug. 2012.
26. L. Lai and H. E. Gamal, "The relay-eavesdropper channel: Cooperation for secrecy," *IEEE Trans. Inf. Theory*, vol. 54, no. 9, pp. 4005–4019, Sept. 2008.
27. Y. Zou, Y.-D. Yao, and B. Zheng, "Opportunistic distributed space-time coding for decode-and-forward cooperation systems," *IEEE Trans. Signal Process.*, vol. 60, no. 4, pp. 1766–1781, Apr. 2012.
28. J. Zhu, Y. Zou, G. Wang, Y.-D. Yao, and G. K. Karagiannidis, "On secrecy performance of antenna selection aided MIMO systems against eavesdropping," *IEEE Trans. Veh. Tech.*, accepted to appear, 2015.
29. Y. Zou, X. Wang, and W. Shen, "Optimal relay selection for physical-layer security in cooperative wireless networks," *IEEE J. Sel. Areas Commun.*, vol. 31, no. 10, pp. 2099–2111, Oct. 2013.
30. Y. Zou, X. Li, and Y.-C. Liang, "Secrecy outage and diversity analysis of cognitive radio systems," *IEEE J. Sel. Areas Commun.*, vol. 32, no. 11, pp. 2222–2236, Nov. 2014.

Chapter 3
Joint Relay and Jammer Selection
for Wireless Physical-Layer Security

Abstract This chapter investigates the physical-layer security for a cooperative wireless network, where a source node transmits to its destination with the aid of multiple intermediate nodes in the face of an eavesdropper. We propose a joint relay and jammer selection scheme for protecting the source-destination transmission against the eavesdropper. To be specific, among the multiple intermediate nodes, one node is first selected to act as the relay for assisting the transmission from the source to destination and meanwhile, another node is chosen as the jammer that is employed to transmit an artificial noise for interfering with the eavesdropper. Also, we consider the conventional pure relay selection and pure jammer selection as benchmark schemes. In the pure relay selection scheme, an intermediate node is selected as the relay to assist the source-destination transmission. By contrast, the pure jammer selection allows an intermediate node to act as the jammer for confusing the eavesdropper. We derive the closed-form intercept probability expressions for the proposed joint relay and jammer selection as well as the conventional pure relay selection and pure jammer selection schemes. Numerical results demonstrate that the joint relay and jammer selection outperforms the pure relay selection and pure jammer selection methods in terms of the intercept probability, showing the security benefit of employing the proposed scheme to protect the wireless communications against eavesdropping.

3.1 System Model and Problem Formulation

As discussed in Chap. 2, given multiple relay nodes available, the "best" node may be chosen and employed to assist the transmission from the source to destination. It has been shown in Chap. 2 that the wireless physical-layer security relying on the relay selection is significantly improved in terms of decreasing the intercept probability. In addition to acting as the relay, a node can also be used to act as the friendly jammer, which emits an artificial noise to confuse the eavesdropper without affecting the legitimate destination. More specifically, as shown in [1–6], an artificial noise can be specially designed in a sophisticated way, so that it only interferes with the eavesdropper, while the legitimate destination is unaffected. In [1], Goel and Negi considered the use of multiple antennas for the artificial noise design and proved that the artificial noise would not interfere with the legitimate destination as

© Springer International Publishing Switzerland 2016   35
Y. Zou, J. Zhu, *Physical-Layer Security for Cooperative Relay Networks*,
Wireless Networks, DOI 10.1007/978-3-319-31174-6_3

long as the number of transmit antennas at the source node is more than the number of receive antennas at the eavesdropper. It was shown in [1–3] that the wireless secrecy capacity is substantially improved by properly using the artificial noise. Additionally, in [4–6], the authors studied the power allocation between the artificial noise and the information-bearing signal of carrying the source information intended to the destination.

In order to take advantage of both the relay and jammer, the joint relay and jammer selection [7–10] is emerging as an effective means of securing the wireless source-destination transmission with the aid of multiple intermediate nodes, where a carefully-selected node acts as the relay for enhancing the reception quality at the legitimate destination, while another node is selected to act as the jammer for imposing interference on the eavesdropper. In [7], the opportunistic selection of two intermediate nodes was examined for enhancing the wireless physical-layer security, where one selected node is used to help the source transmit its signal to the destination and the other node is to generate intentional interference. However, the jamming interference as studied in [7] not only interferes with the eavesdropper, but also adversely affects the legitimate destination. Besides, in [8–10], the joint relay and jammer selection was investigated for two-way cooperative networks under the physical-layer security constraint, where three intermediate nodes are typically selected to participate in defending wireless communications against eavesdropping. Specifically, the first selected node assists the wireless information exchange using the amplify-and-forward (AF) protocol, while the other two nodes are employed to generate intentional interference on the eavesdropper.

In this chapter, we explore the joint relay and jammer selection for enhancing the physical-layer security of a wireless network consisting of a source and a destination with the aid of multiple intermediate nodes in the presence of an eavesdropper. Specifically, an intermediate node is carefully selected to act as the relay for assisting the source-destination transmission by using the decode-and-forward (DF) protocol, differing from [7] in which the AF protocol was considered. Meanwhile, another intermediate node is simultaneously chosen to act as the jammer for transmitting an artificial noise that is specially designed to interfere with the eavesdropper only, which is differing from [8–10], where the intentional interference adversely affects both the destination and eavesdropper. The main contributions of this chapter is summarized as follows. First, we present a joint relay and jammer selection scheme, where two intermediate nodes are chosen to act as the relay and jammer, respectively. For comparison purposes, we also consider the conventional pure relay selection and pure jammer selection as our benchmarks. Second, closed-form intercept probability expressions for the conventional pure relay selection and pure jammer selection as well as the proposed joint relay and jammer selection are derived over Rayleigh fading channels. Additionally, numerical intercept probabilities of these three schemes are provided, showing the security benefit of exploiting the joint relay and jammer selection against eavesdropping.

Fig. 3.1 A source (*S*) transmitting to its destination (*D*) with the aid of *M* intermediate nodes, among which one node is selected as a relay (*R*) to assist the S-D transmission and one node is selected as a jammer (*J*) for interfering with an eavesdropper (*E*)

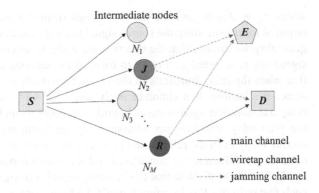

We first provide the system model of a wireless network consisting of a source (*S*) transmitting to a destination (*D*) with the help of *M* intermediate nodes in the presence of an eavesdropper (*E*), as shown in Fig. 3.1. Specifically, among the *M* intermediate nodes, we select one node as a relay for assisting the S-D transmission, where the decode-and-forward (DF) protocol is considered when the relay forwards its received signal from *S* to *D*. Moreover, among the remaining *M* − 1 intermediate nodes, we further select one node to act as a jammer for sending a specially-designed signal (known as the artificial noise) [1–3], which only interferes with *E* in tapping the source signal without affecting the legitimate receiver *D*. For notational convenience, the set of *M* intermediate nodes is represented by $\mathcal{N} = \{N_1, N_2, \cdots, N_M\}$.

As shown in Fig. 3.1, we consider that the *D* and *E* are out of the transmit coverage of *S* and a relay is selected among the *M* intermediate nodes for forwarding the source signal to *D*. When the relay retransmits the source signal, the *E* is assumed to be capable of overhearing the relay transmission. In order to protect the relay transmission against eavesdropping, a jammer node is utilized to send an artificial noise for deteriorating the *E*'s signal reception. Note that all the wireless links between any two nodes of Fig. 3.1 are modeled as independent Rayleigh fading channels. Additionally, a zero-mean additive white Gaussian noise (AWGN) with a variance of N_0 is encountered at any receiver of Fig. 3.1.

Considering that *S* transmits its signal x_s at a power of P_s, we can express the received signal at an intermediate node N_i as

$$y_i = h_{si}\sqrt{P_s}x_s + n_i, \tag{3.1}$$

where h_{si} represents the wireless fading of the S-N_i channel and n_i represents the AWGN received at N_i. Using the Shannon's capacity formula and (3.1), the capacity of the S-N_i channel is given by

$$C_{si} = \frac{1}{2}\log_2(1 + |h_{si}|^2\gamma_s), \tag{3.2}$$

where $\gamma_s = P_s/N_0$ and the factor $\frac{1}{2}$ arises from the fact that two time slots are required for transmitting the source signal to the D via a relay node. Without loss of generality, we assume that the intermediate node N_i succeeds in decoding the source signal and is selected as a relay to forward its decoded outcome. It is pointed out that when the relay transmits to the D, another intermediate node will be selected to act as a jammer that simultaneously sends the artificial noise for protecting the relay transmission against eavesdropping. In order to make a fair comparison with the pure relay transmission without using the jammer, the total transmit power consumed by both the relay R and jammer J is constrained to P_s. For simplicity, the equal power allocation is considered and the transmit powers of R and J are given by $P_s/2$. with the E. Note that the jamming signal is designed in a sophisticated way such that only the E is interfered, while the legitimate receiver D keeps unaffected. Hence, the received signal at D can be expressed as

$$y_d = h_{id}\sqrt{\frac{P_s}{2}}x_s + n_d, \qquad (3.3)$$

where h_{id} represents the wireless fading of the N_i-D channel and n_d represents the AWGN received at N_i. Meanwhile, the E can also overhear the relay transmission and the corresponding overheard signal can be given by

$$y_e = h_{ie}\sqrt{\frac{P_s}{2}}x_s + h_{je}\sqrt{\frac{P_s}{2}}x_n + n_e, \qquad (3.4)$$

where h_{ie} and h_{je} represent the wireless fading of the N_i-E channel and that of the J-E channel, x_n represents the specially-designed artificial noise, and n_e represents the AWGN at E. Using (3.3), we obtain the capacity of the N_i-D channel as

$$C_{id} = \frac{1}{2}\log_2(1 + |h_{id}|^2\frac{\gamma_s}{2}). \qquad (3.5)$$

Similarly, using (3.4), we obtain the capacity of the N_i-E channel as given by

$$C_{ie} = \frac{1}{2}\log_2(1 + \frac{|h_{ie}|^2\gamma_s}{|h_{je}|^2\gamma_s + 2}). \qquad (3.6)$$

So far, we have completed the signal modeling of the S-D transmission with the help of M intermediate nodes. Notice that the wireless magnitudes $|h_{id}|^2$, $|h_{ie}|^2$ and $|h_{je}|^2$ are independent exponential random variables with respective means of σ_{id}^2, σ_{ie}^2 and σ_{je}^2.

3.2   Joint Relay and Jammer Selection for Physical-Layer Security

In this section, we propose a joint relay and jammer selection scheme for wireless physical-layer security and carry out its intercept probability analysis over Rayleigh fading channels. For comparison purposes, we also present the pure relay selection and pure jammer selection methods.

3.2.1   Pure Relay Selection

In this subsection, we first present the conventional pure relay selection as a benchmark. Specifically, S transmits its signal x_s to the M intermediate nodes that attempt to decode the source signal x_s. For notational convenience, we denote the set of nodes that successfully decode x_s by $\mathscr{D}$. Given M intermediate nodes, there are 2^M possible combinations for the decoding set $\mathscr{D}$. Thus, the sample space of D can be given by $\{\emptyset, \mathscr{D}_1, \mathscr{D}_2, \cdots, \mathscr{D}_n, \cdots, \mathscr{D}_{2^M-1}\}$, where $\emptyset$ denotes an empty set and $\mathscr{D}_n$ denotes the n-th non-empty subcollection of the M intermediate nodes. To be specific, the event $\mathscr{D} = \emptyset$ implies that all the M nodes fail to decode x_s, which is described as

$$C_{si} < R, \quad i = 1, 2, \cdots, M, \tag{3.7}$$

where R denotes the data rate and C_{si} is given by (3.2). Meanwhile, the event $\mathscr{D} = \mathscr{D}_n$ means that the nodes within $\mathscr{D}_n$ succeed in decoding x_s and the remaining nodes fail to decode. In an information-theoretic sense, the event $\mathscr{D} = \mathscr{D}_n$ can be described as

$$C_{si} > R, \quad i \in \mathscr{D}_n$$
$$C_{sj} < R, \quad j \in \bar{\mathscr{D}}_n. \tag{3.8}$$

where $\bar{\mathscr{D}}_n = \mathscr{N} - \mathscr{D}_n$ denotes the complementary set of $\mathscr{D}_n$. If the decoding set $\mathscr{D}$ is empty, implying that all the intermediate nodes fail to decode x_s, no node is selected as a relay to forward the source signal. If the decoding set $\mathscr{D}$ is non-empty e.g. $\mathscr{D} = \mathscr{D}_n$, an intermediate node is chosen from the decoding set $\mathscr{D}_n$ for forwarding the source signal x_s to the D. Since the passive E's CSI is difficult to obtain in practice, we consider that only the CSI knowledge of the main channel spanning from the M intermediate nodes to D is known in performing the relay selection. In this case, an intermediate node that successfully decodes x_s and has the highest channel capacity C_{id} is typically selected as the relay to forward the source signal x_s to the D. Thus, from (3.5), the relay selection criterion is expressed as

$$\text{Relay} = \arg \max_{i \in \mathscr{D}_n} C_{id} = \arg \max_{i \in \mathscr{D}_n} |h_{id}|^2. \tag{3.9}$$

Denoting the relay node by 'r', we can obtain the capacity of the channel spanning from the relay to D and that spanning from the relay to E as

$$C_{rd} = \frac{1}{2} \log_2(1 + |h_{rd}|^2 \gamma_s), \tag{3.10}$$

and

$$C_{re} = \frac{1}{2} \log_2(1 + |h_{re}|^2 \gamma_s), \tag{3.11}$$

where $|h_{rd}|^2$ and $|h_{re}|^2$ represent the wireless fading of the channel spanning from the relay to D and that spanning from the relay to E. As aforementioned, an intercept event occurs when the capacity of the main channel C_{rd} falls below that of the wiretap channel C_{re}. Thus, we can obtain the intercept probability of the conventional pure relay selection as

$$P_{\text{int}}^{\text{relay}} = \sum_{n=1}^{2^M-1} \Pr(C_{rd} < C_{re}, \mathscr{D} = \mathscr{D}_n), \tag{3.12}$$

which can be rewritten as

$$P_{\text{int}}^{\text{relay}} = \sum_{n=1}^{2^M-1} \Pr(\mathscr{D} = \mathscr{D}_n) \sum_{i \in \mathscr{D}_n} \Pr(C_{id} < C_{ie}, r = i). \tag{3.13}$$

From (3.8), we can obtain the term $\Pr(\mathscr{D} = \mathscr{D}_n)$ as

$$\Pr(\mathscr{D} = \mathscr{D}_n) = \prod_{i \in \mathscr{D}_n} \exp(-\frac{2\Delta}{\sigma_{si}^2}) \prod_{j \in \bar{\mathscr{D}}_n} [1 - \exp(-\frac{2\Delta}{\sigma_{sj}^2})], \tag{3.14}$$

where $\Delta = \frac{2^{2R}-1}{\gamma_s}$, σ_{si}^2 and σ_{sj}^2 are the means of $|h_{si}|^2$ and $|h_{sj}|^2$, respectively. Moreover, by using (2.32), the term $\Pr(C_{id} < C_{ie}, r = i)$ can be given by

$$\Pr(C_{id} < C_{ie}, r = i) = \frac{\sigma_{ie}^2}{\sigma_{ie}^2 + \sigma_{id}^2} + \sum_{m=1}^{2^{|\mathscr{D}_n|-1}-1} (-1)^{|\mathscr{C}_{n,m}|} (1 + \frac{\sigma_{id}^2}{\sigma_{ie}^2} + \sum_{j \in \mathscr{C}_{n,m}} \frac{\sigma_{id}^2}{\sigma_{jd}^2})^{-1}, \tag{3.15}$$

where σ_{ie}^2, σ_{id}^2 and σ_{jd}^2 are the means of $|h_{ie}|^2$, $|h_{id}|^2$ and $|h_{jd}|^2$, respectively, $\mathscr{C}_{n,m}$ denotes the m-th non-empty collection of the set $\mathscr{D}_n$, and $| \cdot |$ denotes the set cardinality. Substituting $\Pr(\mathscr{D} = \mathscr{D}_n)$ and $\Pr(C_{id} < C_{ie}, r = i)$ from (3.14) and (3.15) into (3.13) gives a closed-form intercept probability expression for the pure relay selection scheme.

3.2.2 Pure Jammer Selection

This subsection presents the pure jammer selection scheme, in which one node is selected among the M intermediate nodes to act as a jammer for interfering with the E. More specifically, when the S transmits its signal x_s to the D, an intermediate node acting as a jammer simultaneously sends an artificial noise for confusing the E without affecting the D. For fair comparisons in terms of power consumption, the total power consumed by the S and J is constrained to P_s. Here, we consider the simple equal power allocation and thus the transmit powers of the S and J are given by $P_s/2$. Note that in the pure jammer selection, the S directly transmits its signal x_s to the D without relying on the relay node. Thus, the received signal at the D can be given by

$$y_d = h_{sd}\sqrt{\frac{P_s}{2}}x_s + n_d, \qquad (3.16)$$

where h_{sd} represents the wireless fading of the S-D channel and n_d represents the AWGN received at D. From (3.16), the capacity of the S-D channel is obtained as

$$C_{sd} = \log_2(1 + |h_{sd}|^2\frac{\gamma_s}{2}). \qquad (3.17)$$

Meanwhile, the S-D transmission can be overheard by the E and the corresponding received signal is expressed as

$$y_e = h_{se}\sqrt{\frac{P_s}{2}}x_s + h_{je}\sqrt{\frac{P_s}{2}}x_n + n_e, \qquad (3.18)$$

where h_{se} and h_{je} represent the wireless fading of the S-E channel and that of the J-E channel, x_n represents the specially-designed artificial noise, and n_e represents the AWGN at E. It needs to be pointed out that the artificial noise x_n is specially designed such that it only interferes with the E without affecting the legitimate receiver D. Interest readers may refer to [1–3] for more details about the artificial noise design. From (3.18), we can obtain the capacity of the S-E channel in the presence of a jammer as

$$C_{se} = \log_2(1 + \frac{|h_{se}|^2\gamma_s}{|h_{je}|^2\gamma_s + 2}). \qquad (3.19)$$

Note that the E is passive and it is challenging to obtain the E's CSI, whereas the J is active and its CSI may be obtained using a channel estimation method [11, 12]. Hence, we assume that the E's CSI is unavailable and the J's CSI is known in carrying out the jammer selection. Under this condition, an intermediate node that can minimize the capacity achieved at the E considered as the jammer for interfering with the E. Hence, from (3.19), the jammer selection criterion is given by

$$\text{Jammer} = \arg\min_{j\in\mathcal{N}} C_{se} = \arg\max_{j\in\mathcal{N}} |h_{je}|^2, \tag{3.20}$$

where $\mathcal{N}$ represents the set of M intermediate nodes. As discussed above, an intercept event happens when the capacity of the S-D channel falls below that of the S-E channel. Hence, by using (3.17), (3.19) and (3.20), the intercept probability of the pure jammer selection scheme is obtained as

$$P_{\text{int}}^{\text{jammer}} = \Pr(C_{sd} < \min_{j\in\mathcal{N}} C_{se}) = \Pr(|h_{sd}|^2 < \frac{2|h_{se}|^2}{\max_{j\in\mathcal{N}} |h_{je}|^2 \gamma_s + 2}), \tag{3.21}$$

which can be rewritten as

$$P_{\text{int}}^{\text{jammer}} = \Pr(\max_{j\in\mathcal{N}} |h_{je}|^2 \gamma_s < \frac{2|h_{se}|^2}{|h_{sd}|^2} - 2). \tag{3.22}$$

Note that $|h_{sd}|^2$ and $|h_{se}|^2$ are independent exponentially distributed random variables with respective means of σ_{sd}^2 and σ_{se}^2. Denoting $X = |h_{sd}|^2$, $Y = |h_{se}|^2$ and $Z = Y/X$, we obtain the cumulative distribution function (CDF) of Z as

$$\Pr(Z < z) = 1 - \int_0^\infty \frac{1}{\sigma_{sd}^2} \exp(-\frac{x}{\sigma_{sd}^2}) dx \int_{xz}^\infty \frac{1}{\sigma_{se}^2} \exp(-\frac{y}{\sigma_{se}^2}) dy$$

$$= 1 - \int_0^\infty \frac{1}{\sigma_{sd}^2} \exp(-\frac{x}{\sigma_{sd}^2} - \frac{xz}{\sigma_{se}^2}) dx$$

$$= \frac{\sigma_{sd}^2 z}{\sigma_{se}^2 + \sigma_{sd}^2 z}, \tag{3.23}$$

from which the probability density function (PDF) of Z is given by

$$p(z) = \frac{\sigma_{sd}^2 \sigma_{se}^2}{(\sigma_{se}^2 + \sigma_{sd}^2 z)^2}. \tag{3.24}$$

Considering that random variables $|h_{je}|^2 (j \in \mathcal{N})$ for different intermediate nodes are independent exponentially distributed and using (3.24), we can obtain (3.22) as

$$P_{\text{int}}^{\text{jammer}} = \Pr(\max_{j\in\mathcal{N}} |h_{je}|^2 \gamma_s < 2z - 2)$$

$$= \int_1^\infty \frac{\sigma_{sd}^2 \sigma_{se}^2}{(\sigma_{se}^2 + \sigma_{sd}^2 z)^2} \prod_{j\in\mathcal{N}} [1 - \exp(-\frac{2z - 2}{\sigma_{je}^2 \gamma_s})] dz, \tag{3.25}$$

where $\mathcal{N}$ represents the set of the M intermediate nodes. By using the binomial expansion theorem, the item $\prod_{j \in \mathcal{N}} [1 - \exp(-\frac{2z-2}{\sigma_{je}^2 \gamma_s})]$ can be given by

$$\prod_{j \in \mathcal{N}} [1 - \exp(-\frac{2z-2}{\sigma_{je}^2 \gamma_s})] = 1 + \sum_{n=1}^{2^M-1} (-1)^{|\mathcal{I}_n|} \exp(-\sum_{j \in \mathcal{I}_n} \frac{2z-2}{\sigma_{je}^2 \gamma_s}), \qquad (3.26)$$

where $\mathcal{I}_n$ denotes the n-th non-empty subcollection of the M intermediate nodes. Substituting (3.26) into (3.25) yields

$$P_{int}^{jammer} = \int_1^\infty \frac{\sigma_{sd}^2 \sigma_{se}^2}{(\sigma_{se}^2 + \sigma_{sd}^2 z)^2} [1 + \sum_{n=1}^{2^M-1} (-1)^{|\mathcal{I}_n|} \exp(-\sum_{j \in \mathcal{I}_n} \frac{2z-2}{\sigma_{je}^2 \gamma_s})] dz$$

$$= \frac{\sigma_{se}^2}{\sigma_{se}^2 + \sigma_{sd}^2} + \sum_{n=1}^{2^M-1} (-1)^{|\mathcal{I}_n|} \Phi(\mathcal{I}_n), \qquad (3.27)$$

where $\Phi(\mathcal{I}_n)$ is given by

$$\Phi(\mathcal{I}_n) = \int_1^\infty \frac{\sigma_{sd}^2 \sigma_{se}^2}{(\sigma_{se}^2 + \sigma_{sd}^2 z)^2} \exp(-\sum_{j \in \mathcal{I}_n} \frac{2z-2}{\sigma_{je}^2 \gamma_s}) dz. \qquad (3.28)$$

Letting $\sigma_{se}^2 + \sigma_{sd}^2 z = x$, we can rewrite (3.28) as

$$\Phi(\mathcal{I}_n) = \exp(\sum_{j \in \mathcal{I}_n} \frac{2\sigma_{sd}^2 + 2\sigma_{se}^2}{\sigma_{sd}^2 \sigma_{je}^2 \gamma_s}) \int_{\sigma_{se}^2 + \sigma_{sd}^2}^\infty \frac{\sigma_{se}^2}{x^2} \exp(-\sum_{j \in \mathcal{I}_n} \frac{2x}{\sigma_{sd}^2 \sigma_{je}^2 \gamma_s}) dx. \qquad (3.29)$$

By substituting $-\sum_{j \in \mathcal{I}_n} \frac{2x}{\sigma_{sd}^2 \sigma_{je}^2 \gamma_s} = t$, (3.29) can be further obtained as

$$\Phi(\mathcal{I}_n) = \sum_{j \in \mathcal{I}_n} \frac{2\sigma_{se}^2}{\sigma_{sd}^2 \sigma_{je}^2 \gamma_s} \exp(C_n) \int_{-\infty}^{-C_n} \frac{\exp(t)}{t^2} dt, \qquad (3.30)$$

where the parameter C_n is given by

$$C_n = \sum_{j \in \mathcal{I}_n} \frac{2(\sigma_{sd}^2 + \sigma_{se}^2)}{\sigma_{sd}^2 \sigma_{je}^2 \gamma_s}. \qquad (3.31)$$

Performing the integration of (3.30) by parts yields

$$\Phi(\mathcal{I}_n) = \sum_{j \in \mathcal{I}_n} \frac{2\sigma_{se}^2}{\sigma_{sd}^2 \sigma_{je}^2 \gamma_s} [C_n^{-1} + Ei(-C_n) \exp(C_n)], \qquad (3.32)$$

where $Ei(x) = \int_{-\infty}^{x} \frac{e^t}{t} dt$ is the exponential integral function. Substituting $\Phi(\mathscr{J}_n)$ from (3.32) into (3.27) gives a close-form expression of the intercept probability for the pure jammer selection scheme.

3.2.3 Joint Relay and Jammer Selection

In this subsection, we propose the joint relay and jammer selection scheme for improving the physical-layer security of *S-D* transmissions with the help of *M* intermediate nodes, among which one node is selected as the *R* for assisting the source transmission and at the same time, another node is chosen as the *J* for confusing the *E*. As discussed above, an intermediate node that succeeds in decoding x_s and maximizes the channel capacity C_{id} is selected as the relay. Meanwhile, another intermediate node that minimizes the *E*'s channel capacity is chosen as the jammer. Hence, from (3.5) and (3.6), the joint relay and jammer selection criterion is written as

$$\text{Relay} = \arg\max_{i \in \mathscr{D}_n} C_{id} = \arg\max_{i \in \mathscr{D}_n} |h_{id}|^2,$$

$$\text{Jammer} = \arg\min_{j \in \mathscr{N}-\{r\}} C_{re} = \arg\max_{j \in \mathscr{N}-\{r\}} |h_{je}|^2, \tag{3.33}$$

where $\mathscr{D}_n$ denotes the decoding set as given by (3.8), r denotes the relay, $\{r\}$ represents the relay set comprised of only one intermediate node that is selected as the *R*, and '$-$' represents the set difference. Similarly to (3.13), the intercept probability of the proposed joint relay and jammer selection scheme is obtained as

$$P_{\text{int}}^{\text{joint}} = \sum_{n=1}^{2^M-1} \Pr(\mathscr{D} = \mathscr{D}_n) \sum_{i \in \mathscr{D}_n} \Pr(C_{id} < C_{ie}, r = i), \tag{3.34}$$

where C_{id} and C_{ie} are given by

$$C_{id} = \frac{1}{2} \log_2(1 + |h_{id}|^2 \frac{\gamma_s}{2}),$$

$$C_{ie} = \frac{1}{2} \log_2(1 + \frac{|h_{ie}|^2 \gamma_s}{\max_{j \in \mathscr{N}-\{i\}} |h_{je}|^2 \gamma_s + 2}), \tag{3.35}$$

where $\{i\}$ represents the set comprised of only one intermediate node N_i. Substituting C_{id} and C_{ie} from (3.35) into (3.34) gives

$$P_{\text{int}}^{\text{joint}} = \sum_{n=1}^{2^M-1} \Pr(\mathscr{D} = \mathscr{D}_n) \sum_{i \in \mathscr{D}_n} \Psi(\mathscr{D}_n), \tag{3.36}$$

where $\Pr(\mathcal{D} = \mathcal{D}_n)$ is given by (3.14) and $\Psi(\mathcal{D}_n)$ is defined as

$$\Psi(\mathcal{D}_n) = \Pr(\max_{j \in \mathcal{N}-\{i\}} |h_{je}|^2 \gamma_s + 2 < 2\frac{|h_{ie}|^2}{|h_{id}|^2}, r = i), \tag{3.37}$$

wherein r denotes the selected relay node. Combining (3.33) and (3.37), we have

$$\Psi(\mathcal{D}_n) = \Pr(\max_{j \in \mathcal{N}-\{i\}} |h_{je}|^2 \gamma_s + 2 < 2\frac{|h_{ie}|^2}{|h_{id}|^2}, \max_{k \in \mathcal{D}_n-\{i\}} |h_{kd}|^2 < |h_{id}|^2). \tag{3.38}$$

Noting that $|h_{id}|^2$ and $|h_{ie}|^2$ are independent exponential random variables and denoting $|h_{id}|^2 = x$ and $|h_{ie}|^2 = y$, we obtain the joint PDF of (x, y) as

$$p(x, y) = \frac{1}{\sigma_{id}^2 \sigma_{ie}^2} \exp(-\frac{x}{\sigma_{id}^2} - \frac{y}{\sigma_{ie}^2}), \tag{3.39}$$

where σ_{id}^2 and σ_{ie}^2 are the expected values of random variables $|h_{id}|^2$ and $|h_{ie}|^2$. Moreover, $|h_{je}|^2$ and $|h_{kd}|^2$ are independent exponentially distributed random variables with respective means of σ_{je}^2 and σ_{kd}^2. Hence, combining (3.38) and (3.39), we obtain

$$\Psi(\mathcal{D}_n) = \Pr(\max_{j \in \mathcal{N}-\{i\}} |h_{je}|^2 \gamma_s + 2 < 2\frac{y}{x}, \max_{k \in \mathcal{D}_n-\{i\}} |h_{kd}|^2 < x)$$

$$= \int_{y>x} \prod_{j \in \mathcal{N}-\{i\}} [1 - \exp(-\frac{2y - 2x}{\sigma_{je}^2 \gamma_s x})] \prod_{k \in \mathcal{D}_n-\{i\}} [1 - \exp(-\frac{x}{\sigma_{kd}^2})] p(x, y) dx dy. \tag{3.40}$$

Using the binomial expansion theorem, we have

$$\prod_{j \in \mathcal{N}-\{i\}} [1 - \exp(-\frac{2y - 2x}{\sigma_{je}^2 \gamma_s x})] = 1 + \sum_{n=1}^{2^{M-1}-1} (-1)^{|\mathcal{G}_n|} \exp(-\sum_{j \in \mathcal{G}_n} \frac{2y - 2x}{\sigma_{je}^2 \gamma_s x}), \tag{3.41}$$

and

$$\prod_{k \in \mathcal{D}_n-\{i\}} [1 - \exp(-\frac{x}{\sigma_{kd}^2})] = 1 + \sum_{m=1}^{2^{|\mathcal{D}_n|-1}-1} (-1)^{|\mathcal{Z}_m|} \exp(-\sum_{k \in \mathcal{Z}_m} \frac{x}{\sigma_{kd}^2}), \tag{3.42}$$

where M is the number of intermediate nodes, and $\mathcal{G}_n$ and $\mathcal{Z}_m$ represent the n-th and m-th non-empty subcollections of the sets $\mathcal{N} - \{i\}$ and $\mathcal{D}_n - \{i\}$, respectively. Substituting $\prod_{j \in \mathcal{N}-\{i\}} [1 - \exp(-\frac{2y-2x}{\sigma_{je}^2 \gamma_s x})]$ and $\prod_{k \in \mathcal{D}_n-\{i\}} [1 - \exp(-\frac{x}{\sigma_{kd}^2})]$ from (3.41) and (3.42) into (3.40) yields

$$\Psi(\mathscr{D}_n) = \Psi_{n1} + \sum_{n=1}^{2^{M-1}-1} (-1)^{|\mathscr{G}_n|} \exp(\sum_{j\in\mathscr{G}_n} \frac{2}{\sigma_{je}^2\gamma_s})\Psi_{n2} + \sum_{m=1}^{2^{|\mathscr{D}_n|-1}-1} (-1)^{|\mathscr{Z}_m|}\Psi_{n3}$$

$$+ \sum_{n=1}^{2^{M-1}-1}\sum_{m=1}^{2^{|\mathscr{D}_n|-1}-1} (-1)^{|\mathscr{G}_n|+|\mathscr{Z}_m|} \exp(\sum_{j\in\mathscr{G}_n} \frac{2}{\sigma_{je}^2\gamma_s})\Psi_{n4},$$

$$\tag{3.43}$$

where Ψ_{n1}, Ψ_{n2}, Ψ_{n3} and Ψ_{n4} are defined as follows

$$\Psi_{n1} = \int_{y>x} p(x,y)dxdy, \tag{3.44}$$

and

$$\Psi_{n2} = \int_{y>x} \exp(-\sum_{j\in\mathscr{G}_n} \frac{2y}{\sigma_{je}^2\gamma_s x})p(x,y)dxdy, \tag{3.45}$$

and

$$\Psi_{n3} = \int_{y>x} \exp(-\sum_{k\in\mathscr{Z}_m} \frac{x}{\sigma_{kd}^2})p(x,y)dxdy, \tag{3.46}$$

and

$$\Psi_{n4} = \int_{y>x} \exp(-\sum_{j\in\mathscr{G}_n} \frac{2y}{\sigma_{je}^2\gamma_s x} - \sum_{k\in\mathscr{Z}_m} \frac{x}{\sigma_{kd}^2})p(x,y)dxdy, \tag{3.47}$$

wherein $p(x,y)$ is given by (3.39). Substituting $p(x,y)$ from (3.39) into (3.44) gives

$$\Psi_{n1} = \int_{y>x} \frac{1}{\sigma_{id}^2\sigma_{ie}^2} \exp(-\frac{x}{\sigma_{id}^2} - \frac{y}{\sigma_{ie}^2})dxdy$$

$$= \int_0^\infty \frac{1}{\sigma_{id}^2} \exp(-\frac{x}{\sigma_{id}^2})dx \int_x^\infty \frac{1}{\sigma_{ie}^2} \exp(-\frac{y}{\sigma_{ie}^2})dy = \frac{\sigma_{ie}^2}{\sigma_{ie}^2 + \sigma_{id}^2}. \tag{3.48}$$

Combining (3.39) and (3.45), we can compute Ψ_{n2} as

$$\Psi_{n2} = \int_{y>x} \exp(-\sum_{j\in\mathscr{G}_n} \frac{2y}{\sigma_{je}^2\gamma_s x})p(x,y)dxdy$$

$$= \int_0^\infty \frac{1}{\sigma_{id}^2} \exp(-\frac{x}{\sigma_{id}^2})dx \int_x^\infty \frac{1}{\sigma_{ie}^2} \exp(-\sum_{j\in\mathscr{G}_n} \frac{2y}{\sigma_{je}^2\gamma_s x} - \frac{y}{\sigma_{ie}^2})dy$$

$$= \frac{1}{\sigma_{id}^2} \exp(-\frac{a_n}{\sigma_{ie}^2}) \int_0^\infty \frac{x}{x+a_n} \exp(-\frac{x}{\sigma_{id}^2} - \frac{x}{\sigma_{ie}^2})dx, \tag{3.49}$$

where $a_n = \sum\limits_{j \in \mathscr{G}_n} \frac{2\sigma_{ie}^2}{\sigma_{je}^2 \gamma_s}$. By denoting $-\frac{x}{\sigma_{id}^2} - \frac{x}{\sigma_{ie}^2} = z$ and $b_i = \frac{1}{\sigma_{id}^2} + \frac{1}{\sigma_{ie}^2}$, (3.49) can be obtained as

$$\Psi_{n2} = \frac{\sigma_{ie}^2}{\sigma_{ie}^2 + \sigma_{id}^2} \exp(-\frac{a_n}{\sigma_{ie}^2}) \int_{-\infty}^{0} \frac{z}{z - a_n b_i} \exp(z) dz. \tag{3.50}$$

Letting $z - a_n b_i = t$, we further obtain Ψ_{n2} from (3.50) as

$$\Psi_{n2} = \frac{\sigma_{ie}^2}{\sigma_{ie}^2 + \sigma_{id}^2} \exp(\frac{a_n}{\sigma_{id}^2}) \int_{-\infty}^{-a_n b_i} (1 + \frac{a_n b_i}{t}) \exp(t) dt$$

$$= \frac{\sigma_{ie}^2}{\sigma_{ie}^2 + \sigma_{id}^2} \exp(\frac{a_n}{\sigma_{id}^2})[\exp(-a_n b_i) + a_n b_i Ei(-a_n b_i)], \tag{3.51}$$

where $Ei(\cdot)$ is the exponential integration. By combining (3.39) and (3.46), Ψ_{n3} can be computed as

$$\Psi_{n3} = \int_{y>x} \exp(-\sum_{k \in \mathscr{Z}_m} \frac{x}{\sigma_{kd}^2}) p(x, y) dx dy$$

$$= \int_{0}^{\infty} \frac{1}{\sigma_{id}^2} \exp(-\frac{x}{\sigma_{id}^2} - \frac{x}{\sigma_{ie}^2} - \sum_{k \in \mathscr{Z}_m} \frac{x}{\sigma_{kd}^2}) dx = (1 + \frac{\sigma_{id}^2}{\sigma_{ie}^2} + \sum_{k \in \mathscr{Z}_m} \frac{\sigma_{id}^2}{\sigma_{kd}^2})^{-1}. \tag{3.52}$$

Additionally, substituting $p(x, y)$ from (3.39) into (3.47), we have

$$\Psi_{n4} = \int_{0}^{\infty} \frac{1}{\sigma_{id}^2} \exp(-\frac{x}{\sigma_{id}^2} - \sum_{k \in \mathscr{Z}_m} \frac{x}{\sigma_{kd}^2}) dx \int_{x}^{\infty} \frac{1}{\sigma_{ie}^2} \exp(-\frac{y}{\sigma_{ie}^2} - \sum_{j \in \mathscr{G}_n} \frac{2y}{\sigma_{je}^2 \gamma_s x}) dy$$

$$= \frac{1}{\sigma_{id}^2} \exp(-\frac{a_n}{\sigma_{ie}^2}) \int_{0}^{\infty} \frac{x}{x + a_n} \exp(-c_k x) dx$$

$$\tag{3.53}$$

where $a_n = \sum\limits_{j \in \mathscr{G}_n} \frac{2\sigma_{ie}^2}{\sigma_{je}^2 \gamma_s}$ and $c_k = \frac{1}{\sigma_{id}^2} + \frac{1}{\sigma_{ie}^2} + \sum\limits_{k \in \mathscr{Z}_m} \frac{1}{\sigma_{kd}^2}$. From (3.53), we obtain

$$\Psi_{n4} = \frac{1}{\sigma_{id}^2} \exp(-\frac{a_n}{\sigma_{ie}^2}) \int_{0}^{\infty} (1 - \frac{a_n}{x + a_n}) \exp(-c_k x) dx$$

$$= \frac{1}{\sigma_{id}^2 c_k} \exp(-\frac{a_n}{\sigma_{ie}^2}) - \frac{1}{\sigma_{id}^2} \exp(-\frac{a_n}{\sigma_{ie}^2}) \int_{0}^{\infty} \frac{a_n}{x + a_n} \exp(-c_k x) dx. \tag{3.54}$$

By denoting $x + a_n = -\frac{t}{c_k}$ and substituting $x = -\frac{t}{c_k} - a_n$ into (3.54), we arrive at

$$\Psi_{n4} = \frac{1}{\sigma_{id}^2 c_k} \exp(-\frac{a_n}{\sigma_{ie}^2}) + \frac{a_n}{\sigma_{id}^2} \exp(-\frac{a_n}{\sigma_{ie}^2} + c_k a_n) Ei(-a_n c_k). \qquad (3.55)$$

Finally, combining Ψ_{n1}, Ψ_{n2}, Ψ_{n3} and Ψ_{n4} with (3.43) determines the parameter $\Psi(\mathcal{D}_n)$, which is substituted into (3.36) to obtain a closed-form expression of intercept probability for the proposed joint relay and jammer selection scheme.

3.3 Numerical Results and Discussions

In this section, we present numerical intercept probability results of the conventional pure relay selection, pure jammer selection as well as the proposed joint relay and jammer selection schemes. Specifically, the numerical intercept probabilities of these three schemes are obtained by computing (3.12), (3.21) and (3.36). In the numerical evaluation, the average gain of the main channel is specified to $\sigma_{sd}^2 = \sigma_{si}^2 = \sigma_{id}^2 = 1$ and the average gain of the wiretap channel is given by $\sigma_{ie}^2 = \sigma_{je}^2 = 0.2$, unless otherwise mentioned. In addition, a transmission rate of $R = 1$ bit/s/Hz and $M = 5$ are used, unless otherwise stated.

Figure 3.2 shows the intercept probability comparison among the conventional pure relay selection, pure jammer selection as well as the proposed joint relay and jammer selection schemes by plotting (3.12), (3.21) and (3.36) as a function of SNR γ_s. We also provide the simulated intercept probability results for these three scheme in Fig. 3.2, where the continuous lines and discrete markers represent the theoretical and simulated intercept probabilities, respectively. It is shown from Fig. 3.2 that as the SNR γ_s increases, the intercept probabilities of the pure jammer selection and the proposed joint relay and jammer selection schemes are reduced significantly. By contrast, with an increasing SNR, the intercept probability of the pure relay selection converges to a constant, showing that an intercept probability floor occurs for the pure relay selection in high SNR region.

One can also see from Fig. 3.2 that as the SNR increases, the pure jammer selection initially performs worse and finally becomes better than the pure relay selection in terms of its intercept probability. Figure 3.2 also shows that the intercept probability of the proposed joint relay and jammer selection scheme is strictly less than that of the conventional pure relay selection and pure jammer selection methods. This confirms the security advantage of the proposed joint relay and jammer selection over the conventional pure relay selection and pure jammer selection. Additionally, it can be observed from Fig. 3.2 that the theoretical and simulated intercept probability results match each other very well, showing the correctness of our intercept probability analysis.

In Fig. 3.3, we depict the intercept probability versus the SNR γ of the proposed joint relay and jammer selection for different data rates R. As shown in Fig. 3.3, as the data rate R increases from $R = 0.6$ to 1 bit/s/Hz, the intercept probability of the

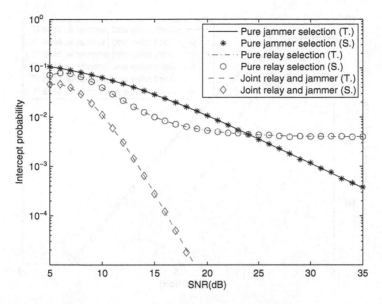

Fig. 3.2 Intercept probability versus SNR γ_s of the conventional pure relay selection, pure jammer selection as well as the proposed joint relay and jammer selection schemes

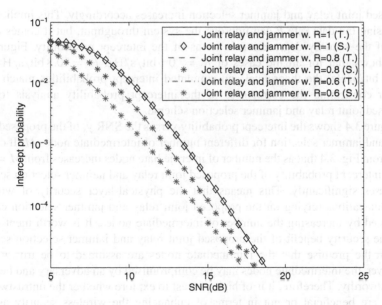

Fig. 3.3 Intercept probability versus SNR γ_s of the proposed joint relay and jammer selection scheme for different data rates R

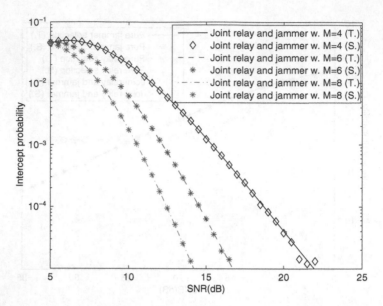

Fig. 3.4 Intercept probability versus SNR γ_s of the proposed joint relay and jammer selection scheme for different number of intermediate nodes M

proposed joint relay and jammer selection increases accordingly. This implies that increasing the data rate R can improve the system throughput, but it comes at the cost of the security degradation in terms of the intercept probability. Figure 3.3 also shows that for all the cases of $R = 0.6\,\text{bit/s/Hz}$, $R = 0.8\,\text{bit/s/Hz}$ and $R = 1\,\text{bit/s/Hz}$, the theoretical and simulated intercept probabilities match well, further confirming the correctness of the intercept probability analysis for the proposed joint relay and jammer selection scheme.

Figure 3.4 shows the intercept probability versus the SNR γ_s of the proposed joint relay and jammer selection for different number of intermediate nodes M. It can be seen from Fig. 3.4 that as the number of intermediate nodes increases from $M = 4$ to 8, the intercept probability of the proposed joint relay and jammer selection scheme decreases significantly. This means that the physical-layer security of wireless communications relying on the proposed joint relay and jammer selection can be enhanced by increasing the number of intermediate nodes. It is worth mentioning that the security benefit of the proposed joint relay and jammer selection scheme lies on the premise that the intermediate nodes are assumed to be trustworthy. However, the intermediate nodes may be compromised by an adversary and become untrustworthy. Therefore, it is of high interest to explore whether the untrustworthy relays are beneficial or not in terms of enhancing the wireless security against eavesdropping.

3.4 Conclusions

In this chapter, we have investigated the joint relay and jammer selection for a wireless network consisting of a source and a destination with the help of multiple intermediate nodes against eavesdropping. To be specific, in the joint relay and jammer selection process, an intermediate node is selected and used to act as the relay for assisting the transmission from the source to destination and at the same time, another intermediate node is chosen and employed to act as the jammer for generating an artificial noise, which is specially designed to interfere with the eavesdropper only without affecting the legitimate destination. For comparison purposes, the conventional pure relay selection and pure jammer selection have also been considered as benchmark schemes. We have derived closed-form intercept probability expressions for the proposed joint relay and jammer selection as well as the conventional pure relay selection and pure jammer selection over Rayleigh fading channels. It has been shown that the proposed joint relay and jammer selection strictly performs better than both the pure relay selection and pure jammer selection in terms of the intercept probability. Additionally, with an increasing number of intermediate nodes, the intercept probability of the proposed joint relay and jammer selection is reduced significantly.

References

1. S. Goel and R. Negi, "Guaranteeing secrecy using artificial noise," *IEEE Trans. Wirel. Commun.*, vol. 7, no. 6, pp. 2180–2189, Jul. 2008.
2. D. Goeckel, *et al.*, "Artificial noise generation from cooperative relays for everlasting secrecy in two-hop wireless networks," *IEEE J Sel. Areas Commun.*, vol. 29, no. 10, pp. 2067–2076, Oct. 2011.
3. A. Khisti and D. Zhang, "Artificial-noise alignment for secure multicast using multiple antennas," *IEEE Commun. Lett.*, vol. 17, no. 8, pp. 1568–1571, Aug. 2013.
4. X. Zhou and M. McKay, "Secure transmission with artificial noise over fading channels: Achievable rate and optimal power allocation," *IEEE Trans. Veh. Tech.*, vol. 59, no. 8, pp. 3831–3842, Aug. 2010.
5. N. Romero-Zurita, M. Ghogho, and D. McLernon, "Outage probability based power distribution between data and artificial noise for physical layer security," *IEEE Signal Process. Lett.*, vol. 19, no. 2, pp. 71–74, Feb. 2012.
6. Y. Yang, *et al.*, "Transmitter beamforming and artificial noise with delayed feedback: Secrecy rate and power allocation," *J. Commun. and Net.*, vol. 14, no. 4, pp. 374–384, Aug. 2012.
7. I. Krikidis, J. S. Thompson, and S. McLaughlin, "Relay selection for secure cooperative networks with jamming," *IEEE Trans. Wireless Commun.*, vol. 8, no. 10, pp. 5003–5011, Oct. 2009.
8. J. Chen, *et al.*, "Joint relay and jammer selection for secure two-way relay networks," in *Proc. 2011 IEEE Intern. Commun.*, Kyoto Japan, Jun. 2011.
9. J. Chen, *et al.*, "Joint relay and jammer selection for secure two-way relay networks," *IEEE Trans. Inf. Foren. Sec.*, vol. 7, no. 1, pp. 310–320, Feb. 2012.

10. R. Zhang, *et al.*, "Physical layer security for two-way untrusted relaying with friendly jammers," *IEEE Trans. Veh. Tech.*, vol. 61, no. 8, pp. 3693–3704, Aug. 2012.
11. O. Edfors, *et al.*, "OFDM channel estimation by singular value decomposition," *IEEE Trans. Commun.*, vol. 46, no. 7, pp. 931–939, Jul. 1998.
12. S. Coleri, *et al.*, "Channel estimation techniques based on pilot arrangement in OFDM systems," *IEEE Trans. Broadcasting*, vol. 48, no. 3, pp. 223–229, Mar. 2002.

Chapter 4
Security-Reliability Tradeoff for Cooperative Relay Networks

Abstract This chapter investigates the security-reliability tradeoff (SRT) for wireless communications in the face of multiple eavesdroppers, where the security and reliability are measured by using the intercept probability encountered at the eavesdroppers and the outage probability experienced by the legitimate destination, respectively. We consider a cooperative relay network consisting of a source and a destination with the aid of multiple relay nodes and present two relay selection schemes, namely the single-relay selection (SRS) and multi-relay selection (MRS), to protect the security of source-destination transmissions against eavesdropping. To be specific, in the SRS scheme, only the single "best" relay is selected to assist the source-destination transmissions, whereas in the MRS scheme, multiple relays are enabled to participate in forwarding the source signal to the destination. For comparison purposes, we also consider the traditional direct transmission as a benchmark. We carry out the SRT analysis for the direct transmission as well as the SRS and MRS schemes in terms of deriving their closed-form intercept and outage probability expressions over Rayleigh fading channels. Numerical results demonstrate that both the SRS and MRS schemes perform better than the conventional direct transmission in terms of their SRTs, showing the security and reliability benefits of exploiting the relay selection against eavesdropping. Additionally, with an increasing number of eavesdroppers, the SRTs of the SRS and MRS schemes degrade. By contrast, as the number of relay nodes increases, the SRTs of both the SRS and MRS schemes improve significantly and the MRS consistently outperforms the SRS in terms of their SRTs.

4.1 System Model and Problem Formulation

Recently, an increasing research attention has been paid to improve the secrecy capacity of wireless communications in fading environments [1–3]. The multi-input multi-output (MIMO) was studied extensively and well for enhancing the wireless secrecy capacity by relying on the transmit antenna selection [4–8], artificial noise [9–13], and beamforming techniques [14–16]. To be specific, the transmit antenna selection [8] allows the use of a specific antenna with the highest instantaneous secrecy rate for transmitting the source signal to the destination, which is capable of improving the wireless secrecy capacity. The artificial noise [9] can be designed

© Springer International Publishing Switzerland 2016
Y. Zou, J. Zhu, *Physical-Layer Security for Cooperative Relay Networks*,
Wireless Networks, DOI 10.1007/978-3-319-31174-6_4

along with the MIMO structure in a sophisticated way so that it only interferes with the eavesdropper without affecting the legitimate destination, thus leading to a secrecy capacity improvement. In addition, the MIMO beamforming [14] enables the source node to transmit its signal in a specific direction, so that the source signal received at the legitimate destination encountering constructive interference and becomes much stronger than that overheard by the eavesdropper which experiences destructive interference.

The aforementioned research efforts are devoted to enhance the wireless physical-layer security without considering the communication reliability. As discussed in [17], the low probability of intercept can be achieved by constraining the eavesdropper's decoding error rate. For instance, when the capacity of the wiretap channel falls below that of the main channel, the source may adjust its data rate between the capacity of the main channel and that of the wiretap channels for depriving the eavesdropper with arbitrarily low decoding error rate while reliably communicating to the destination. To elaborate a little further, according to the Shannon coding theorem [18], when the capacity of wiretap channel drops below the data rate, it is impossible for the eavesdropper to succeed in decoding the source signal. However, if the capacity of wiretap channel is higher than the data rate, the eavesdropper becomes capable of successfully decoding the source signal and an intercept event happens in this case. Although decreasing the transmit power can reduce the capacity of the wiretap channel and leads to an increased intercept probability, it comes at the expense of wireless reliability degradation since a lower transmit power decreases the capacity of the main channel and the outage probability experienced by the legitimate destination increases.

To this end, in [19], we investigated the security-reliability tradeoff (SRT) for wireless communications, where the security is quantified by the probability that the eavesdropper successfully intercepts the source signal and the reliability is measured by the probability that an outage event occurs at the legitimate destination, termed as the intercept probability and outage probability, respectively. It was shown in [19] that upon increasing the intercept probability, the outage probability is decreased and vice versa, implying a tradeoff between the security and reliability. In [19], we also proposed an opportunistic relaying selection scheme for the sake of enhancing the SRT and showed that the wireless SRT significantly improves with an increasing number of relays. Furthermore, in [20], we proposed a multi-relay selection scheme, which is shown to achieve a better SRT performance than the opportunistic relaying selection of [20].

In this chapter, we investigate the security-reliability tradeoff for a cooperative relay network in the presence of multiple eavesdroppers, differing from [19] and [20], where only one eavesdropper was considered. The main contributions of this chapter can be summarized as follows. First, we present the single-relay selection (SRS) and multi-relay selection (MRS) schemes, where the SRS chooses only the single "best" relay to assist the source-destination transmission, while in the MRS scheme, multiple relays are selected to participate in forwarding the source transmission to the destination. Second, we derive closed-form expressions of the intercept probability and outage probability for both the SRS and MRS schemes

Fig. 4.1 A cooperative relay
network comprised of one
source (S) transmitting to its
destination (D) with the help
of M relay nodes in the face
of N eavesdroppers (E)

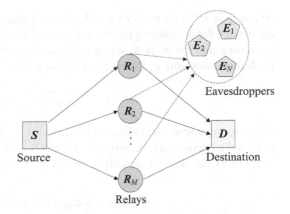

in the face of multiple eavesdroppers. Finally, numerical results show that as the number of eavesdroppers increases, the SRTs of both the SRS and MRS schemes degrade, which can be compensated by exploiting more relay nodes.

We first present the system model of a cooperative relay network in the presence of N eavesdroppers. As shown in Fig. 4.1, a source (S) intends to transmit its signal to a destination (D), however the D is assumed to be out of the S' transmit coverage. To this end, M relay nodes are exploited to assist the S-D transmission and the decode-and-forward (DF) protocol is considered at the relay nodes. For notational convenience, the set of M DF relays is represented by $\boldsymbol{R} = \{R_1, R_2, \cdots, R_M\}$.

When the relay nodes forward the source signal to the D, the N eavesdroppers are considered to be capable of overhearing the relay transmissions and may succeed in decoding the source signal. Following the physical-layer security literature, the eavesdroppers are assumed to know everything about the S-D transmission, except that the source signal is confidential. For notational convenience, the set of N eavesdroppers is denoted by $\boldsymbol{E} = \{E_1, E_2, \cdots, E_N\}$. Additionally, we consider that all the wireless links between any two network nodes of Fig. 4.1 are modeled as independent Rayleigh fading channels. Moreover, a zero-mean AWGN with a variance of N_0 encounters at any receiver. Considering that the S transmits its signal denoted by x_s at a power of P_s and a data rate of R, we can express the received signal at a relay node R_i as

$$y_i = h_{si}\sqrt{P_s}x_s + n_i, \tag{4.1}$$

where h_{si} represents the fading coefficient of the wireless channel spanning from S to R_i and n_i represents the zero mean AWGN with a variance of N_0 received at R_i. Using (4.1), we obtain the channel capacity achieved at R_i as

$$C_{si} = \frac{1}{2}\log_2(1 + |h_{si}|^2\gamma_s), \tag{4.2}$$

where $\gamma_s = P_s/N_0$ is the SNR and the factor $\frac{1}{2}$ aries from the fact that two time slots are needed for completing the transmission of the source signal to the D via a relay. Assuming that the relay R_i successfully decodes the source signal and retransmits its decoded outcome x_s, the received signal at D is written as

$$y_d = h_{id} \sqrt{P_s} x_s + n_d, \tag{4.3}$$

where h_{id} represents the fading coefficient of the wireless channel spanning from R_i to D and n_d represents the zero-mean AWGN with a variance of N_0 received at D. From (4.3), the channel capacity achieved at D is given by

$$C_{id} = \frac{1}{2} \log_2(1 + |h_{id}|^2 \gamma_s). \tag{4.4}$$

Meanwhile, the broadcast nature of the radio propagation leads to the fact that the relay transmission may be overheard by an eavesdropper denoted by E_k. Thus, the received signal at E_k can be given by

$$y_{e_k} = h_{ie_k} \sqrt{P_s} x_s + n_{e_k}, \tag{4.5}$$

where h_{ie_k} represents the fading coefficient of the wireless channel spanning from R_i to e_k and n_{e_k} represents the AWGN received at E_k. Using (4.4), we obtain the channel capacity achieved at the E_k as

$$C_{ie_k} = \frac{1}{2} \log_2(1 + |h_{ie_k}|^2 \gamma_s). \tag{4.6}$$

Throughout this chapter, we assume that the N eavesdroppers are uncoordinated and perform the signal interception independently of each other. Hence, the overall channel capacity achieved at the N eavesdroppers is considered to be the maximum of individual achievable rates of the N eavesdroppers, yielding

$$C_{ie} = \max_{k \in \mathscr{E}} C_{ie_k} = \max_{k \in \mathscr{E}} \frac{1}{2} \log_2(1 + |h_{ie_k}|^2 \gamma_s), \tag{4.7}$$

where $\mathscr{E}$ denotes the set of the N eavesdroppers. As discussed above, if C_{id} falls below C_{ie}, it is impossible to achieve the perfect secrecy. In other words, the source transmission becomes insecure and an intercept event happens in this case. We are motivated to explore the relay selection for decreasing the intercept probability of the S-D transmission with the aid of M relays in the face of N eavesdroppers.

4.2 Security-Reliability Tradeoff for Relay Selection

In this section, we present two specific relay selection schemes, namely the SRS and MRS, and analyze the SRT of the two schemes over Rayleigh fading channels. To be specific, the wireless security is measured by the probability that the eavesdroppers succeed in intercepting the source transmission, referred to as the intercept probability. By contrast, the reliability is characterized by using the probability that an outage event occurs at the destination, called outage probability. Following [19] and [20], we first recall the definition of the outage probability.

Definition 4.1. Letting C_d denote the channel capacity achieved at the destination, we express the outage probability as

$$P_{\text{out}} = \Pr(C_d < R), \tag{4.8}$$

where R represents the data rate. *Additionally, the intercept probability is defend as follows.*

Definition 4.2. Letting C_e denote the channel capacity achieved at the eavesdroppers, the intercept probability is given by

$$P_{\text{int}} = \Pr(C_e > R). \tag{4.9}$$

4.2.1 Direct Transmission

For performance comparisons, we first present the direct transmission as a benchmark. In the direct transmission, the S transmits its message x_s directly to the D without relying on any relay nodes. Meanwhile, the eavesdroppers are assumed to be capable of overhearing the source transmission. Considering that S transmits x_s with the power of P_s, the received signal at D is expressed as

$$y_d = h_{sd}\sqrt{P_s}x_s + n_d, \tag{4.10}$$

where h_{sd} represents the fading coefficient of the channel from S to D and n_d represents the AWGN received at D. Meanwhile, the source transmission is assumed to be overheard by the eavesdroppers. Hence, the received signal at E_k is written as

$$y_{e_k} = h_{se_k}\sqrt{P_s}x_s + n_{e_k}, \tag{4.11}$$

where h_{se_k} represents the fading coefficient of the channel from S to E_k. From (4.10), we obtain the capacity of the S-D channel as

$$C_{sd} = \log_2(1 + |h_{sd}|^2\gamma_s). \tag{4.12}$$

By using (4.11), the channel capacity achieved at the E_k relying on the direct transmission scheme is given by

$$C_{se_k} = \log_2(1 + |h_{se_k}|^2 \gamma_s). \tag{4.13}$$

Considering that the N eavesdroppers are independently of each other, the overall channel capacity achieved at the N eavesdroppers relying on the direct transmission is expressed as

$$C_{se} = \max_{k \in \mathscr{E}} C_{se_k} = \max_{k \in \mathscr{E}} \log_2(1 + |h_{se_k}|^2 \gamma_s), \tag{4.14}$$

where $\mathscr{E}$ represents the set of the N eavesdroppers. From the outage definition as given by (4.8), we obtain the outage probability of the direct transmission scheme as

$$P_{\text{out}}^{\text{direct}} = \Pr(C_{sd} < R), \tag{4.15}$$

where R represents the data rate. Substituting C_{sd} from (4.12) into (4.15) gives

$$P_{\text{out}}^{\text{direct}} = \Pr(|h_{sd}|^2 < \Lambda), \tag{4.16}$$

where $\Lambda = (2^R - 1)/\gamma_s$. Since the random variable $|h_{sd}|^2$ is exponentially distributed, we have

$$P_{\text{out}}^{\text{direct}} = 1 - \exp(-\frac{\Lambda}{\sigma_{sd}^2}), \tag{4.17}$$

where σ_{sd}^2 is the mean of $|h_{sd}|^2$. Additionally, using the intercept definition as given by (4.9), we obtain the intercept probability of the direct transmission as

$$P_{\text{int}}^{\text{direct}} = \Pr(C_{se} > R). \tag{4.18}$$

Substituting C_{se} from (4.14) into (4.19) yields

$$P_{\text{int}}^{\text{direct}} = \Pr(\max_{k \in \mathscr{E}} |h_{se_k}|^2 > \Lambda), \tag{4.19}$$

where $\Lambda = (2^R - 1)/\gamma_s$. Noting that random variables $|h_{se_k}|^2$ for different eavesdroppers e_k are independent of each other, we have

$$P_{\text{int}}^{\text{direct}} = \prod_{k=1}^{N} \exp(-\frac{\Lambda}{\sigma_{se_k}^2}) = \exp(-\sum_{k=1}^{N} \frac{\Lambda}{\sigma_{se_k}^2}), \tag{4.20}$$

where N is the number of eavesdroppers. Combining (4.17) and (4.20), we obtain the relationship between the intercept probability and outage probability as

$$P_{\text{out}}^{\text{direct}} = 1 - (P_{\text{int}}^{\text{direct}})^{(\sum_{k=1}^{N} \frac{\sigma_{sd}^2}{\sigma_{sek}^2})^{-1}}.$$ (4.21)

It can be observed from (4.21) that with an increase of the intercept probability, the outage probability decreases accordingly and vice versa, showing a tradeoff between the security and reliability, namely the security-reliability tradeoff (SRT). Also, one can see from (4.21) that the SRT performance is independent of the SNR γ_s, implying that increasing the transmit power P_s can not achieve any SRT improvements.

4.2.2   Single-Relay Selection

This subsection presents the single-relay selection scheme, where only the single best relay node may be invoked for assisting the S-D transmission. To be specific, S first transmits its source signal x_s to the M relay nodes, which then decode their received signals for recovering x_s. For notational convenience, let $\mathscr{D}$ represent the set of relay nodes that succeed in decoding x_s, called the *decoding set*. Given M relay nodes, there are 2^M possible combinations for the decoding set $\mathscr{D}$ and the corresponding sample space of $\mathscr{D}$ can be expressed as $\Omega = \{\emptyset, \mathscr{D}_1, \mathscr{D}_2, \cdots, \mathscr{D}_m, \cdots, \mathscr{D}_{2^M-1}\}$, where $\emptyset$ represents a null set and $\mathscr{D}_n$ represents the m-th non-empty subset of the M relay nodes. According to the Shannon's coding theorem, a relay node is deemed to fail to decode the source signal, if its channel capacity C_{si} drops below the data rate R. Otherwise, the relay would be able to succeed in decoding the source signal. Thus, by using (4.2), the event $\mathscr{D} = \emptyset$ can be described as

$$C_{si} < R, \quad i = 1, 2, \cdots, M,$$ (4.22)

where C_{si} is given by (4.2). Meanwhile, the event $\mathscr{D} = \mathscr{D}m$ is expressed as

$$\begin{aligned} C_{si} > R, \quad & i \in \mathscr{D}_m \\ C_{sj} < R, \quad & j \in \bar{\mathscr{D}}_m, \end{aligned}$$ (4.23)

where $\bar{\mathscr{D}}_m$ denotes the complementary set of $\mathscr{D}_m$. If the decoding set $\mathscr{D}$ is null, meaning that no relay successfully decodes x_s, all relay nodes remain silent and transmit nothing. In this case, both the D and eavesdroppers become incapable of decoding the source signal x_s. If the decoding set is non-empty, a single relay would be selected from the decoding set $\mathscr{D}$ to forward its decoded signal x_s to the D. In general, a relay node that maximizes the channel capacity C_{id} is selected and used for forwarding the source signal x_s to the D. Hence, given $\mathscr{D} = \mathscr{D}_m$ and using (4.4), we obtain the relay selection criterion as

$$\text{Best Relay} = \arg\max_{i \in \mathcal{D}_m} C_{id} = \arg\max_{i \in \mathcal{D}_m} |h_{id}|^2, \tag{4.24}$$

from which only the CSI $|h_{id}|^2$ is assumed without requiring the eavesdroppers' CSI knowledge $|h_{ie_k}|^2$. Combining (4.4) and (4.24), we obtain the channel capacity from the "best" relay to the D as

$$C_{bd} = \max_{i \in \mathcal{D}_m} \frac{1}{2} \log_2(1 + |h_{id}|^2 \gamma_s), \tag{4.25}$$

where the subscript 'b' denotes the selected "best" relay. Meanwhile, from (4.7), the channel capacity from the "best" relay to the M independent eavesdroppers as

$$C_{be} = \max_{k \in \mathcal{E}} \frac{1}{2} \log_2(1 + |h_{be_k}|^2 \gamma_s). \tag{4.26}$$

Using the law of total probability and the outage definition as given by (4.8), we obtain the outage probability of the SRS scheme as

$$P_{\text{out}}^{\text{SRS}} = \Pr(C_{bd} < R, \mathcal{D} = \emptyset) + \sum_{m=1}^{2^M - 1} \Pr(C_{bd} < R, \mathcal{D} = \mathcal{D}_m), \tag{4.27}$$

where C_{bd} is given by (4.25). Given $\mathcal{D} = \emptyset$, all relays fail to decode the source signal x_s and no relay would be chosen to forward x_s, leading to $C_{bd} = 0$ for $\mathcal{D} = \emptyset$. Substituting this result into (4.27) yields

$$P_{\text{out}}^{\text{SRS}} = \Pr(\mathcal{D} = \emptyset) + \sum_{m=1}^{2^M - 1} \Pr(C_{bd} < R, \mathcal{D} = \mathcal{D}_m). \tag{4.28}$$

Using (4.22), (4.23) and (4.25), we rewrite (4.28) as

$$P_{\text{out}}^{\text{SRS}} = \prod_{i=1}^{M} \Pr(|h_{si}|^2 < \Delta) + \sum_{m=1}^{2^M - 1} \prod_{i \in \mathcal{D}_m} \Pr(|h_{si}|^2 > \Delta) \prod_{j \in \bar{\mathcal{D}}_m} \Pr(|h_{sj}|^2 < \Delta)$$
$$\times \Pr(\max_{i \in \mathcal{D}_m} |h_{id}|^2 < \Delta), \tag{4.29}$$

where $\Delta = (2^{2R} - 1)/\gamma_s$. Noting that random variables $|h_{si}|^2$ and $|h_{id}|^2$ are independent and exponentially distributed, we obtain

$$\Pr(|h_{si}|^2 < \Delta) = 1 - \exp(-\frac{\Delta}{\sigma_{si}^2}), \tag{4.30}$$

and

$$\Pr(\max_{i \in \mathscr{D}_m} |h_{id}|^2 < \Delta) = \prod_{i \in \mathscr{D}_m} \left[1 - \exp(-\frac{\Delta}{\sigma_{id}^2})\right], \qquad (4.31)$$

where σ_{si}^2 and σ_{id}^2 are the means of $|h_{si}|^2$ and $|h_{id}|^2$, respectively. From the intercept definition, the intercept probability of the SRS scheme is obtained as

$$P_{\text{int}}^{\text{SRS}} = \Pr(C_{be} > R, \mathscr{D} = \emptyset) + \sum_{m=1}^{2^M-1} \Pr(C_{be} > R, \mathscr{D} = \mathscr{D}_m), \qquad (4.32)$$

where C_{be} is given by (4.26). In the case of $\mathscr{D} = \emptyset$, no relay would be selected to forward the source transmission and all the eavesdroppers become incapable of tapping the source signal, since the eavesdroppers are assumed to be beyond the S' coverage. Thus, we have $C_{be} = 0$ for $\mathscr{D} = \emptyset$. Substituting this result into (4.32) yields

$$P_{\text{int}}^{\text{SRS}} = \sum_{m=1}^{2^M-1} \prod_{i \in \mathscr{D}_m} \Pr(|h_{si}|^2 > \Delta) \prod_{j \in \bar{\mathscr{D}}_m} \Pr(|h_{sj}|^2 < \Delta) \Pr(\max_{k \in \mathscr{E}} |h_{be_k}|^2 > \Delta).$$

$$(4.33)$$

Noting that random variables $|h_{si}|^2$ and $|h_{sj}|^2$ are exponentially distributed with respective means of σ_{si}^2 and σ_{sj}^2, we have

$$\Pr(|h_{si}|^2 > \Delta) = \exp(-\frac{\Delta}{\sigma_{si}^2}), \qquad (4.34)$$

and

$$\Pr(|h_{sj}|^2 < \Delta) = 1 - \exp(-\frac{\Delta}{\sigma_{sj}^2}). \qquad (4.35)$$

According to (4.24), the single "best" relay node may be selected among the decoding set $\mathscr{D}_m$. Hence, by using the law of total probability, the term $\Pr(\max_{k \in \mathscr{E}} |h_{be_k}|^2 > \Delta)$ is obtained as

$$\Pr(\max_{k \in \mathscr{E}} |h_{be_k}|^2 > \Delta) = \sum_{i \in \mathscr{D}_m} \Pr(\max_{k \in \mathscr{E}} |h_{ie_k}|^2 > \Delta, b = i)$$

$$= \sum_{i \in \mathscr{D}_m} \Pr(\max_{k \in \mathscr{E}} |h_{ie_k}|^2 > \Delta, |h_{id}|^2 > \max_{j \in \mathscr{D}_m - \{i\}} |h_{jd}|^2)$$

$$= \sum_{i \in \mathscr{D}_m} \Pr(\max_{k \in \mathscr{E}} |h_{ie_k}|^2 > \Delta) \Pr(\max_{j \in \mathscr{D}_m - \{i\}} |h_{jd}|^2 < |h_{id}|^2), \qquad (4.36)$$

where the second equality arises by using (4.24) and $\mathscr{D}_m - \{i\}$ represents the set difference between the decoding set $\mathscr{D}_m$ and $\{i\}$. Since the random variables $|h_{ie_k}|^2$ for different eavesdroppers are independent exponentially distributed, we have

$$\Pr(\max_{k \in \mathscr{E}} |h_{ie_k}|^2 > \Delta) = 1 - \prod_{k=1}^{N} [1 - \exp(-\frac{\Delta}{\sigma_{ie_k}^2})], \tag{4.37}$$

where N is the number of eavesdroppers. Moreover, noting that $|h_{jd}|^2$ and $|h_{id}|^2$ are independent exponentially distributed random variables and denoting $|h_{jd}|^2 = x_j$ and $|h_{id}|^2 = y$, we obtain

$$\Pr(\max_{j \in \mathscr{D}_m - \{i\}} |h_{jd}|^2 < |h_{id}|^2) = \int_0^\infty \frac{1}{\sigma_{id}^2} \exp(-\frac{y}{\sigma_{id}^2}) \prod_{j \in \mathscr{D}_m - \{i\}} [1 - \exp(-\frac{y}{\sigma_{jd}^2})] dy, \tag{4.38}$$

where σ_{jd}^2 and σ_{id}^2 are the means of $|h_{jd}|^2$ and $|h_{id}|^2$, respectively. By using the binomial expansion theorem, the term $\prod_{j \in \mathscr{D}_m - \{i\}} [1 - \exp(-\frac{y}{\sigma_{jd}^2})]$ can be written as

$$\prod_{j \in \mathscr{D}_m - \{i\}} [1 - \exp(-\frac{y}{\sigma_{jd}^2})] = 1 + \sum_{m=1}^{2^{|\mathscr{D}_m|-1}-1} (-1)^{|\mathscr{C}_n(m)|} \exp(-\sum_{j \in \mathscr{C}_n(m)} \frac{y}{\sigma_{jd}^2}), \tag{4.39}$$

where $\mathscr{C}_n(m)$ denotes the m-th non-empty subset of $\mathscr{D}_n - \{i\}$ and $|\mathscr{C}_n(m)|$ denotes the cardinality of $\mathscr{C}_n(m)$. Combining (4.38) and (4.39), we arrive at

$$\Pr(\max_{j \in \mathscr{D}_m - \{i\}} |h_{jd}|^2 < |h_{id}|^2) = 1 + \sum_{m=1}^{2^{|\mathscr{D}_m|-1}-1} (-1)^{|\mathscr{C}_n(m)|} (1 + \sum_{j \in \mathscr{C}_n(m)} \frac{\sigma_{id}^2}{\sigma_{jd}^2})^{-1} \tag{4.40}$$

Combining (4.37) and (4.40) with (4.36) determines the term $\Pr(\max_{k \in \mathscr{E}} |h_{be_k}|^2 > \Delta)$, which is then substituted into (4.33) to obtain a closed-form intercept probability expression for the SRS scheme. So far, we have derived the closed-form expressions of the outage probability and intercept probability for the SRS scheme over Rayleigh fading channels.

4.2.3 Multi-relay Selection

This subsection presents the multi-relay selection scheme. To be specific, given a non-empty decoding set $\mathscr{D} = \mathscr{D}_m$, all the relay nodes within $\mathscr{D}_m$ are used or simultaneously transmitting the source signal x_s to the D. By contrast, in the SRS scheme, only the single "best" relay is selected among the decoding set to forward x_s. In the MRS scheme, a weight vector denoted by $\mathbf{w} = [w_1, w_2, \cdots, w_{|\mathscr{D}_m|}]^T$ is utilized by all the relay nodes of $\mathscr{D}_m$ in transmitting the source signal x_s, where $|\mathscr{D}_m|$ denotes the cardinality of $\mathscr{D}_m$. In order to make a fair comparison with the SRS scheme, the total transmit power of all the relay nodes is constrained to P_s. Thus, considering that all the relay nodes within the decoding set $\mathscr{D}_m$ simultaneously transmit x_s with the aid of a weight vector $\mathbf{w}$, we can express the received signal at the D as

$$y_d^{\text{MRS}} = \sqrt{P_s}\mathbf{w}^T\mathbf{h}_d x_s + n_d, \tag{4.41}$$

where $\mathbf{h}_d = [h_{1d}, h_{2d}, \cdots, h_{|\mathscr{D}_m|d}]^T$. Meanwhile, all the M eavesdroppers also overhear the relays' transmissions and the signal received at an eavesdropper E_k is written as

$$y_e^{\text{MRS}} = \sqrt{P_s}\mathbf{w}^T\mathbf{h}_{e_k} x_s + n_e, \tag{4.42}$$

where $\mathbf{h}_e = [h_{1e_k}, h_{2e_k}, \cdots, h_{|\mathscr{D}_m|e_k}]^T$. From (4.41), the received SNR at D is expressed as

$$\text{SNR}_d^{\text{MRS}} = |\mathbf{w}^T\mathbf{h}_d|^2\gamma_s. \tag{4.43}$$

Similarly, using (4.42), we obtain the received SNR at E_k is expressed as

$$\text{SNR}_{e_k}^{\text{MRS}} = |\mathbf{w}^T\mathbf{h}_{e_k}|^2\gamma_s. \tag{4.44}$$

In general, the passive eavesdroppers' CSIs are unknown and thus the weight vector $\mathbf{w}$ is optimized to maximize $\text{SNR}_d^{\text{MRS}}$ without considering $\text{SNR}_{e_k}^{\text{MRS}}$, yielding

$$\max_{\mathbf{w}} \text{SNR}_d^{\text{MRS}} = \max_{\mathbf{w}} |\mathbf{w}^T\mathbf{h}_d|^2, \quad \text{s.t. } ||\mathbf{w}|| = 1, \tag{4.45}$$

where the constraint $||\mathbf{w}|| = 1$ arises for the normalization purposes. By using the Cauchy-Schwarz inequality, the optimal weight vector $\mathbf{w}_{\text{opt}}$ can be obtained from (4.45) as

$$\mathbf{w}_{\text{opt}} = \frac{\mathbf{h}_d^*}{|\mathbf{h}_d|}, \tag{4.46}$$

where the optimal weight vector $\mathbf{w}_{\text{opt}}$ only requires the CSI of the main channel $\mathbf{h}_d$ without requiring the eavesdroppers' CSI $\mathbf{h}_{e_k}$. Substituting $\mathbf{w}_{\text{opt}}$ from (4.46) into (4.43), we obtain the channel capacity achieved at the D as

$$C_d^{\text{MRS}} = \frac{1}{2}\log_2(1 + \sum_{i \in \mathscr{D}_m} |h_{id}|^2\gamma_s). \tag{4.47}$$

Meanwhile, substituting $\mathbf{w}_{\text{opt}}$ from (4.46) into (4.44), we obtain the eavesdropper E_k's channel capacity as

$$C_{e_k}^{\text{MRS}} = \frac{1}{2} \log_2(1 + \frac{|\mathbf{h}_d^H \mathbf{h}_{e_k}|^2}{|\mathbf{h}_d|^2} \gamma_s), \tag{4.48}$$

where H represents the Hermitian transpose. Considering that the M eavesdroppers are independent and uncoordinated in tapping the source signal, the overall capacity of the wiretap channel is given by

$$C_e^{\text{MRS}} = \max_{k \in \mathscr{E}} \frac{1}{2} \log_2(1 + \frac{|\mathbf{h}_d^H \mathbf{h}_{e_k}|^2}{|\mathbf{h}_d|^2} \gamma_s), \tag{4.49}$$

where $\mathscr{E}$ represents the set of M eavesdroppers. In what follows, we analyze the outage probability and intercept probability of the MRS scheme over Rayleigh fading channels. According to the outage definition as given by (4.8), the outage probability of the MRS can be obtained as

$$P_{\text{out}}^{\text{MRS}} = \Pr(\mathscr{D} = \emptyset) + \sum_{m=1}^{2^M-1} \Pr(C_d^{\text{MRS}} < R, \mathscr{D} = \mathscr{D}_m). \tag{4.50}$$

Using (4.22), (4.23) and (4.47), we rewrite (4.50) as

$$P_{\text{out}}^{\text{MRS}} = \prod_{i=1}^{N} \Pr(|h_{si}|^2 < \Delta) + \sum_{m=1}^{2^M-1} \prod_{i \in \mathscr{D}_m} \Pr(|h_{si}|^2 > \Delta) \prod_{j \in \bar{\mathscr{D}}_m} \Pr(|h_{sj}|^2 < \Delta)$$
$$\times \Pr(\sum_{i \in \mathscr{D}_m} |h_{id}|^2 < \Delta), \tag{4.51}$$

where the terms $\Pr(|h_{si}|^2 < \Delta)$, $\Pr(|h_{si}|^2 > \Delta)$ and $\Pr(|h_{sj}|^2 < \Delta)$ can be given by

$$\Pr(|h_{si}|^2 < \Delta) = 1 - \exp(-\frac{\Delta}{\sigma_{si}^2})$$

$$\Pr(|h_{si}|^2 > \Delta) = \exp(-\frac{\Delta}{\sigma_{si}^2}) \tag{4.52}$$

$$\Pr(|h_{sj}|^2 < \Delta) = 1 - \exp(-\frac{\Delta}{\sigma_{sj}^2}).$$

However, obtaining a closed-form expression for the term $\Pr(\sum_{i \in \mathscr{D}_m} |h_{id}|^2 < \Delta)$ is difficult. For simplicity, we here assume that the random variables $|h_{id}|^2$ for different relay nodes are independent and identically distributed (i.i.d.) with an average channel gain as denoted by $\sigma_d^2 = E(|h_{id}|^2)$. This assumption is valid in a statistical sense when all the relay nodes are uniformly distributed over a certain geographical area, which is commonly used in the open literature [21–23]. Considering that the random variables $|h_{id}|^2$ for $i \in \mathscr{D}_m$ are i.i.d., we have

$$\Pr(\sum_{i \in \mathscr{D}_m} |h_{id}|^2 < \Delta) = \Gamma(\frac{\Delta}{\sigma_d^2}, |\mathscr{D}_m|), \qquad (4.53)$$

where $\Gamma(x, k)$ is known as the incomplete Gamma function, which is defined as

$$\Gamma(x, k) = \int_0^x \frac{t^{k-1}}{\gamma_s(k)} e^{-t} dt. \qquad (4.54)$$

Additionally, using the intercept definition as given by (4.2), we obtain the intercept probability of the MRS scheme as

$$P_{\text{int}}^{\text{MRS}} = \sum_{m=1}^{2^M-1} \prod_{i \in \mathscr{D}_m} \Pr(|h_{si}|^2 > \Delta) \prod_{j \in \bar{\mathscr{D}}_m} \Pr(|h_{sj}|^2 < \Delta) \Pr(\max_{k \in \mathscr{E}} \frac{|\mathbf{h}_d^H \mathbf{h}_{ek}|^2}{|\mathbf{h}_d|^2} > \Delta), \qquad (4.55)$$

where the terms $\Pr(|h_{si}|^2 > \Delta)$ and $\Pr(|h_{sj}|^2 < \Delta)$ can be readily determined in closed-form. However, obtaining a closed-form expression for $\Pr(\max_{k \in \mathscr{E}} \frac{|\mathbf{h}_d^H \mathbf{h}_{ek}|^2}{|\mathbf{h}_d|^2} > \Delta)$ is challenging. Nevertheless, we can evaluate the intercept probability of the MRS scheme with aid of computer simulations.

4.3 Numerical Results and Discussions

This section presents the numerical SRT comparisons among the direct transmission, SRS and MRS schemes. To be specific, the SRT performance of the direct transmission, SRS and MRS schemes are characterized in terms of their outage probabilities and intercept probabilities by evaluating (4.17), (4.20), (4.29), (4.33), (4.51) and (4.55). In the numerical evaluation, the average gains of the main channel and wiretap channels are given by $\sigma_{sd}^2 = \sigma_{si}^2 = \sigma_{id}^2 = 1$ and $\sigma_{ie_k}^2 = 0.1$. In addition, an SNR of $\gamma_s = 10\,\text{dB}$ and a data rate of $R = 1$ bit/s/Hz are utilized, unless otherwise stated.

Figure 4.2 illustrates the outage probability versus the SNR γ_s of the direct transmission, SRS and MRS schemes with $M = 6$ and $N = 2$, where M and N represent the number of relays and eavesdroppers, respectively. One can observe from Fig. 4.2 that with an increasing SNR γ_s, the outage probabilities of the direct transmission, SRS and MRS schemes are reduced, showing that increasing the transmit power indeed improves the wireless reliability by reducing

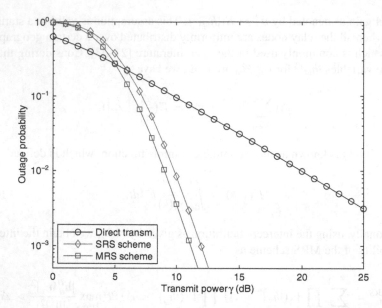

Fig. 4.2 Outage probability versus the SNR γ_s of the direct transmission, SRS and MRS schemes

the outage probability. Figure 4.2 also shows that in high SNR region, the direct transmission performs the worst and the MRS scheme is the best in terms of their outage probabilities.

Figure 4.3 shows the intercept probability versus the SNR γ_s of the direct transmission, SRS and MRS schemes with $M = 6$ and $N = 2$. One can observe from Fig. 4.3 that with an increasing SNR γ_s, the intercept probabilities of the direct transmission, SRS and MRS schemes increase accordingly. This means that increasing the transmit power degrades the wireless security with an increased intercept probability, although it improves the wireless reliability with a reduced outage probability. One can also see from Fig. 4.3 that the intercept probabilities of the SRS and MRS schemes are less than that of the direct transmission, showing the security benefits of the proposed SRS and MRS.

Figure 4.4 shows the outage probability versus the data rate R of the direct transmission, SRS and MRS schemes with $M = 6$ and $N = 2$. It can be seen from Fig. 4.4 that as the data rate R increases, the outage performance of the direct transmission, SRS and MRS schemes degrades accordingly. This is because that with an increased data rate, it becomes more likely that the channel capacity of the source-destination transmission falls below the data rate. Figure 4.4 also shows that the SRS and MRS schemes perform better than the direct transmission in terms of the outage probability in low data rate region. Moreover, as the data rate increases, the outage performance of proposed SRS and MRS schemes may become worse than that of the direct transmission. This is due to the fact that in the SRS and MRS schemes, the half-duplex relaying constraint is considered, which sacrifices the spectrum efficiency as compared to the direct transmission.

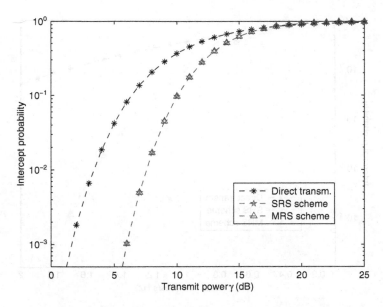

Fig. 4.3 Intercept probability versus the SNR γ_s of the direct transmission, SRS and MRS schemes

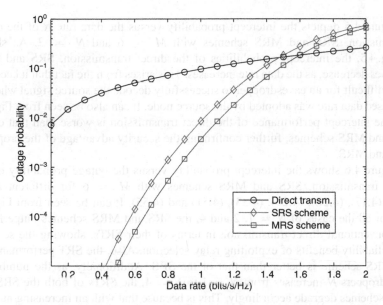

Fig. 4.4 Outage probability versus the data rate R of the direct transmission, SRS and MRS schemes

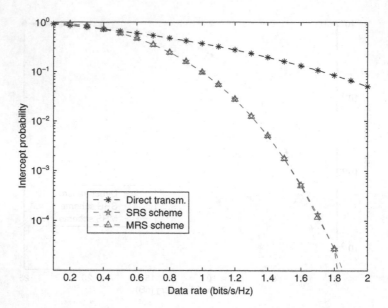

Fig. 4.5 Intercept probability versus the data rate R of the direct transmission, SRS and MRS schemes

Figure 4.5 depicts the intercept probability versus the data rate R of the direct transmission, SRS and MRS schemes with $M = 6$ and $N = 2$. As shown in Fig. 4.5, the intercept probabilities of the direct transmission, SRS and MRS schemes decrease, as the data rate increases. This arises from the fact that it becomes more difficult for an eavesdropper to successfully decode the source signal when an increased data rate was adopted by the source node. It can also be seen from Fig. 4.5 that the intercept performance of the direct transmission is worse than that of the SRS and MRS schemes, further confirming the security advantage of the proposed SRS and MRS.

Figure 4.6 shows the intercept probability versus the outage probability of the direct transmission, SRS and MRS schemes with $M = 6$ for different N by using (4.17), (4.20), (4.29), (4.33), (4.51) and (4.55). It can be seen from Fig. 4.6 that for all the cases of $N = 1$, 2 and 4, the SRS and MRS schemes outperforms the conventional direct transmission in terms of their SRTs, showing the security and reliability benefits of exploiting relay selection. Also, the SRT performance of the MRS scheme is better than that of the SRS. Additionally, as the number of eavesdroppers N increases from $N = 1$ to $N = 4$, the SRTs of both the SRS and MRS schemes degrade accordingly. This is because that with an increasing number of eavesdroppers, it becomes more likely that an eavesdropper succeeds in decoding the source signal.

In Fig. 4.7, we demonstrate the intercept probability versus the outage probability of the direct transmission, SRS and MRS schemes with $N = 2$ for different M. One can see from Fig. 4.7 that for all the cases of $M = 2$, 4 and 8, the SRS and MRS

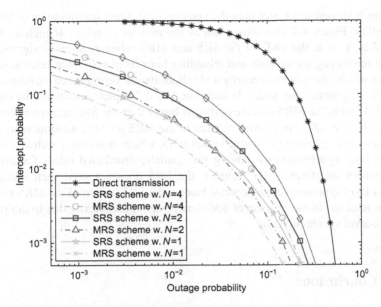

Fig. 4.6 SRT comparisons among the direct transmission, SRS and MRS schemes for different N, where N represents the number of eavesdroppers

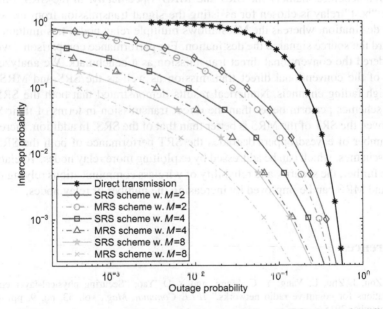

Fig. 4.7 SRT comparisons among the direct transmission, SRS and MRS schemes for different N, where N represents the number of relay nodes

schemes both perform better than the conventional direct transmission in terms of their SRTs. Figure 4.7 also shows that as the number of relays M increases from $M = 2$ to $N = 8$, the SRTs of the SRS and MRS schemes improve significantly, further confirming the security and reliability benefits of exploiting relay selection. In other words, the wireless security and reliability can be improved simultaneously by exploiting more relay nodes. In addition, Fig. 4.7 depicts that for all the cases of $N = 2$, 4 and 8, the MRS scheme performs better than the SRS scheme in terms of their SRTs. It needs to be pointed out that in the MRS scheme, multiple relay nodes simultaneously transmit the source signal to D, which, however, requires a precise symbol-level synchronization among the spatially-distributed relays for avoiding inter-symbol interference. By contrast, the SRS does not need such a complex symbol-level synchronization process. Hence, the SRT benefit of the MRS over the SRS is achieved at the expense of additional system complexity due to the precise symbol-level synchronization.

4.4   Conclusions

In this chapter, we investigated the security-reliability tradeoff for wireless relay networks in the presence of multiple eavesdroppers and presented two relay selection schemes, namely the SRS and MRS. Specifically, in the SRS, only the single "best" relay is chosen for assisting the signal transmission from the source to the destination, whereas the MRS allows multiple relay nodes to simultaneously forward the source signal to the destination. For performance comparisons, we also considered the conventional direct transmission as a benchmark. We analyzed the SRTs of the conventional direct transmission as well as the SRS and MRS over Rayleigh fading channels. Numerical results demonstrated that both the SRS and MRS schemes perform better than the direct transmission in terms of their SRTs. Moreover, the SRT of the MRS is better than that of the SRS. In addition, increasing the number of eavesdroppers degrades the SRT performance of both the SRS and MRS schemes, which can be addressed by exploiting more relay nodes. To elaborate a little further, the security and reliability of wireless communications relying on the SRS and MRS can be improved by increasing the number of relay nodes.

References

1. Y. Zou, J. Zhu, L. Yang, Y.-.C. Liang, and Y.-D. Yao, "Securing physical-layer communications for cognitive radio networks," *IEEE Commun. Mag.*, vol. 53, no. 9, pp. 48–54, September 2015.
2. W. Trappe, "The challenges facing physical layer security," *IEEE Commun. Mag.*, vol. 53, no. 6, pp. 16–20, June 2015.
3. A. Mukherjee, S. A. Fakoorian, J. Huang, and A. L. Swindlehurst, "Principles of physical layer security in multiuser wireless networks: A survey," *IEEE Commun. Surveys & Tutorials*, vol. 16, no. 3, pp. 1550–1573, September 2014.

4. H. Alves, R. D. Souza, M. Debbah, and M. Bennis, "Performance of transmit antenna selection physical layer security schemes," *IEEE Signal Process. Lett.*, vol. 19, no. 6, pp. 372–375, Jun. 2012.
5. N. S. Ferdinand, D. B. da Costa and M. Latva-Aho, "Effects of outdated CSI on the secrecy performance of MISO wiretap channels with transmit antenna selection," *IEEE Commun. Lett.*, vol. 17, no. 5, pp. 864–867, May 2013.
6. S. Yan, N. Yang, R. Malaney, and J. Yuan, "Transmit antenna selection with Alamouti coding and power allocation in MIMO wiretap channels," *IEEE Trans. Wireless Commun.*, vol. 13, no. 3, pp. 1656–1667, Mar. 2014.
7. J. Zhu, Y. Zou, G. Wang, Y.-D. Yao, and G. K. Karagiannidis, "On secrecy performance of antenna selection aided MIMO systems against eavesdropping," *IEEE Trans. Veh. Tech.*, accepted to appear.
8. Y. Zou, J. Zhu, X. Wang, and V. Leung, "Improving physical-layer security in wireless communications through diversity techniques," *IEEE Network*, vol. 29, no. 1, pp. 42–48, Jan. 2015.
9. A. Ozcelikkale and T. M. Duman, "Cooperative precoding and artificial noise design for security over interference channels," *IEEE Sig. Process. Lett.*, vol. 22, no. 12, pp. 2234–2238, Dec. 2015.
10. N. Romero-Zurita, M. Ghogho, and D. McLernon, "Outage probability based power distribution between data and artificial noise for physical layer security," *IEEE Sig. Process. Lett.*, vol. 19, no. 2, pp. 71–74, Feb. 2012.
11. D. Goeckel, *et al.*, "Artificial noise generation from cooperative relays for everlasting secrecy in two-hop wireless networks," *IEEE J. Select. Areas Commun.*, vol. 29, no. 10, pp. 2067–2076, Oct. 2010.
12. S. H. Chae, *et al.*, "Enhanced secrecy in stochastic wireless networks: Artificial noise with secrecy protected zone," *IEEE Trans. Infor. Forensics and Security*, vol. 9, no. 10, pp. 1617–1628, Oct. 2014.
13. A. Khisti and D. Zhang, "Artificial-noise alignment for secure multicast using multiple antennas," *IEEE Commun. Lett.*, vol. 17, no. 8, pp. 1568–1571, Aug. 2013.
14. A. Mukherjee, "Robust beamforming for security in MIMO wiretap channels with imperfect CSI," *IEEE Trans. Signal Process.*, vol. 59, no. 1, pp. 351–361, Jan. 2011.
15. J. Zhang and M. C. Gursoy, "Relay beamforming strategies for physical-layer security," in *Proceedings of the 44th Annual Conference on Information Sciences and Systems (2010 IEEE CISS)*, Princeton, NJ, USA, Mar. 2010.
16. S. Bashar, Z. Ding, and G. Y. Li, "On secrecy of codebook-based transmission beamforming under receiver limited feedback," *IEEE Trans. Wirel. Commun.*, vol. 10, no. 4, pp. 1212–1223, Apr. 2011.
17. A. O. Hero, "Secure space-time communication," *IEEE Trans. Inform. Theory*, vol. 49, no. 12, pp. 3235–3249, Dec. 2003.
18. C. E. Shannon, "A mathematical theory of communication," *Bell System Techn. J.*, vol. 27, pp. 379–423, 1948.
19. Y. Zou, X. Wang, W. Shen, and L. Hanzo, "Security versus reliability analysis of opportunistic relaying," *IEEE Trans. Veh. Tech.*, vol. 63, no. 6, pp. 2653–2661, Jul. 2014.
20. Y. Zou, B. Champagne, W.-P. Zhu, and L. Hanzo, "Relay-selection improves the security-reliability trade-off in cognitive radio systems," *IEEE Trans. Commun.*, vol. 63, no. 1, pp. 215–228, Jan. 2015.
21. I. Krikidis, J. S. Thompson, and S. McLaughlin, "Relay selection for secure cooperative networks with jamming," *IEEE Trans. Wirel. Commun.*, vol. 8, no. 10, pp. 5003–5011, Oct. 2009.
22. T. Nechiporenko, *et al.*, "On the capacity of Rayleigh fading cooperative systems under adaptive transmission," *IEEE Trnas. Wirel. Commun.*, vol. 8, no. 4, pp. 1626–1631, Apr. 2009.
23. G. Farhadi and N. C. Beaulieu, "On the performance of amplify-and-forward cooperative systems with fixed gain relays," *IEEE Trans. Wirel. Commun.*, vol. 7, no. 5, pp. 1851–1856, May 2008.

Chapter 5
Improving Security-Reliability Tradeoff Through Joint Relay and Jammer Selection

Abstract This chapter explores the employment of joint relay and jammer selection for improving the security-reliability tradeoff (SRT) of wireless transmission from a source to a destination with the help of multiple relays and jammers in the presence of an eavesdropper. We present a joint relay and jammer selection scheme for protecting the source transmission against eavesdropping. To be specific, a relay is selected among multiple relay candidates to assist the source transmission to the destination, while a friendly jammer is employed to emit the artificial noise for preventing the eavesdropper from decoding the source transmission. For comparison purposes, the conventional pure relay selection and pure jammer selection are considered as benchmark schemes. We derive closed-form expressions of both the intercept probability and outage probability for the conventional pure relay selection and pure jammer selection as well as the proposed joint relay and jammer selection schemes over Rayleigh fading channels. Numerical results illustrate that the SRT performance of proposed joint relay and jammer selection is better than that of the conventional pure relay selection and pure jammer selection, showing the SRT enhancement of employing the relay and jammer selection. It is also shown that with an increasing number of relays and jammers, the SRT of wireless communications relying on the proposed joint relay and jammer selection is improved significantly.

5.1 System Model and Problem Formulation

In Chap. 4, we discussed the tradeoff between security and reliability, called security-reliability tradeoff (SRT) [1–3], for wireless communications and showed that the intercept probability of wireless communications can be reduced at the cost of an increasing outage probability and vice versa. We also examined the use of relay nodes in Chap. 4 in order to improve the wireless SRT performance, which was mainly focused on the design of relay selection for enhancing the wireless reliability without jointly considering how to prevent an eavesdropper from intercepting the source transmission. The relay selection was first investigated in [4] to combat the fading effect for wireless reliability improvement and shown to be capable of achieving the full diversity gain. Recently, the use of relay selection for wireless reliability was studied extensively from different perspectives in terms of e.g. the bit error rate (BER), outage probability, and coverage extension. Since there exists a

© Springer International Publishing Switzerland 2016

Y. Zou, J. Zhu, *Physical-Layer Security for Cooperative Relay Networks*,
Wireless Networks, DOI 10.1007/978-3-319-31174-6_5

tradeoff between the security and reliability, the wireless reliability benefit achieved by using the relay selection can also be converted to the security improvement [5–9].

As previously discussed in Chap. 3, a friendly jammer may be invoked to emit the artificial noise, which is designed onto the null space of the main channel (spanning from the source to destination) so that the legitimate destination is unaffected [10–12]. Due to the fact that the wiretap channel (spanning from the source to eavesdropper) is independent of the main channel, the specially-designed artificial noise is not necessarily in the null space of the wiretap channel and would thus cause a certain amount of interference to the eavesdropper. It was shown in Chap. 3 that the wireless physical-layer security can be improved notably by exploiting a friendly jammer with the aid of the artificial noise in terms of decreasing the intercept probability. In Chap. 3, the joint use of the relay and friendly jammer was studied in the context of wireless communications against both the eavesdropping and fading effects. It was shown that the joint relay and jammer selection is better than the pure relay selection and pure jammer selection in terms of protecting the wireless physical-layer security. Notice that the joint relay and jammer selection was examined from the security aspect only without considering the wireless reliability issue.

To this end, this chapter investigates the joint relay and jammer selection from the SRT perspective, which takes into account both the security and reliability, differing from the chapter, where only the wireless security was addressed. We consider a wireless network consisting of a source and a destination with the help of N relay nodes and M friendly jammers in the presence of an eavesdropper. The main contributions of this chapter is summarized as follows. First, we present a joint relay and jammer selection scheme, where a relay node is selected among N candidates to assist the source transmission to the destination and meanwhile, a friendly jammer is chosen among M jammers to transmit the artificial noise for confusing the eavesdropping and preventing it from decoding the source transmission. For comparison purposes, the conventional pure relay selection and pure jammer selection are also considered as benchmark schemes. Second, closed-form expressions of both the intercept probability and outage probability are derived for the conventional pure relay selection and pure jammer selection as well as the proposed joint relay and jammer selection over Rayleigh fading channels. In addition, through the numerical SRT evaluation, it is shown that the proposed joint relay and jammer selection scheme performs better than the conventional pure relay selection and pure jammer selection in terms of their SRTs. Moreover, the SRT performance of proposed joint relay and jammer selection can be improved significantly by exploiting more number of relay nodes and friendly jammers.

We first presents the system model of a wireless network that consists of a source (S) transmitting to its intended destination (D) with the help of N relay nodes and M friendly jammers in the presence of an eavesdropper (E), as shown in Fig. 5.1. To be specific, a relay node is selected among the N candidates to help the source transmit to the destination, where the DF protocol is considered when the selected relay forwards its received signal from the S to D. Meanwhile, a jammer is chosen among the M friendly jammers and employed to transmit the artificial noise, which

Fig. 5.1 A source (S) transmits to its intended destination (D) with the help of N relays and M friendly jammers in the presence of an eavesdropper (E)

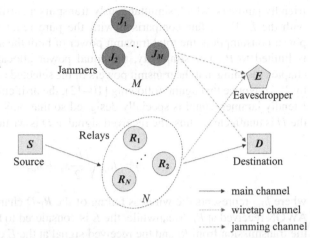

is specially designed [1–3] for interfering with the E only without affecting the D. For notational convenience, the sets of N relays and M jammers are denoted by $\mathcal{R} = \{R_1, R_2, \cdots, R_N\}$ and $\mathcal{J} = \{J_1, J_2, \cdots, J_M\}$, respectively.

As depicted in Fig. 5.1, we consider that both the D and E are beyond the transmit coverage of S and a relay is selected among the N relay candidates to help the source transmit its signal to the D. When the selected relay retransmits the source signal, the E is assumed to be capable of overhearing the relay transmission. In order to protect the relay transmission against eavesdropping, a jammer is chosen among the M friendly jammers to send an artificial noise for interfering with the E. It needs to be pointed that all the wireless links between any two nodes of Fig. 5.1 are modeled as independent Rayleigh fading channels. Additionally, a zero-mean AWGN with a variance of N_0 is encountered at any receiver of Fig. 5.1.

Considering that the S transmits its signal x_s at a power of P_s, we can express the received signal at a relay node R_i as

$$y_i = h_{sr_i}\sqrt{P_s}x_s + n_i, \tag{5.1}$$

where h_{sr_i} represents the wireless fading of S-R_i channel and n_i represents the AWGN received at R_i. From (5.1), the capacity of the S-R_i channel is given by

$$C_{sr_i} = \frac{1}{2}\log_2(1 + |h_{sr_i}|^2\gamma_s), \tag{5.2}$$

where $\gamma_s = P_s/N_0$ and the factor $\frac{1}{2}$ arises from the fact that two time slots are required for transmitting the source signal to the D via a relay node. Without loss of generality, we consider that a relay R_i of successfully decoding the source signal is selected to forward its decoded outcome to the D. Again, when the selected relay retransmits to the D, a jamming node denoted by J_m is chosen among the M

friendly jammers, which simultaneously transmits an artificial noise for interfering with the E. For a fair comparison with the pure relay selection in terms of the power consumption, the total transmit power of both the selected relay and jammer is limited to P_s. For simplicity, the equal power allocation is considered in the chapter, meaning that the transmit powers of the selected relay and jammer are given by $P_s/2$. Notice that again, following [10–12], the artificial noise transmitted by the friendly jammer signal is specially designed so that only the E is interfered, while the D is unaffected. Thus, the received signal at D is written as

$$y_d = h_{r_id}\sqrt{\frac{P_s}{2}}x_s + n_d, \tag{5.3}$$

where h_{r_id} represents the wireless fading of the R_i-D channel and n_d represents the AWGN received at R_i. Meanwhile, the E is considered to be capable of overhearing the transmission from R_i and the received signal at the E can be expressed as

$$y_e = h_{r_ie}\sqrt{\frac{P_s}{2}}x_s + h_{j_me}\sqrt{\frac{P_s}{2}}x_n + n_e, \tag{5.4}$$

where h_{r_ie} and h_{j_me} represent the wireless fading of the R_i-E channel and that of the J_m-E channel, x_n represents the artificial noise, and n_e represents the AWGN received at the E. From (5.3), we obtain the capacity of the R_i-D channel as

$$C_{r_id} = \frac{1}{2}\log_2(1 + |h_{r_id}|^2\frac{\gamma_s}{2}). \tag{5.5}$$

Similarly, with the help of a friendly jammer j_m, we obtain the capacity of the R_i-E channel from (5.4) as

$$C_{r_ie}(j_m) = \frac{1}{2}\log_2(1 + \frac{|h_{r_ie}|^2\gamma_s}{|h_{j_me}|^2\gamma_s + 2}). \tag{5.6}$$

which completes the signal modeling of the source-destination transmission with the help of N relay nodes and M friendly jammers. Since all the wireless links (i.e., h_{r_id}, h_{r_ie} and h_{j_me}) are modeled as independent Rayleigh fading channels, one can readily obtain that the fading magnitudes $|h_{r_id}|^2$, $|h_{r_ie}|^2$ and $|h_{j_me}|^2$ are independent exponentially distributed, whose means are denoted by $\sigma_{r_id}^2$, $\sigma_{r_ie}^2$ and $\sigma_{j_me}^2$, respectively.

5.2 Security-Reliability Tradeoff for Joint Relay and Jammer Selection

In this section, we propose the joint relay and jammer selection scheme for enhancing the wireless SRT performance. For comparison purposes, we also present the conventional pure relay selection and pure jammer selection as benchmarks.

5.2.1 Pure Relay Selection

Let us first consider the conventional pure relay selection, where a single "best" relay node is selected among the N candidates to assist the S-D transmission. To be specific, the S first transmits x_s to the N relay nodes, which then attempt to decode their received signals. For notational convenience, these relay nodes that successfully decode the source signal x_s are represented by a set $\mathscr{D}$, called the decoding set. Given N relay nodes, there are 2^N possible combinations for the set $\mathscr{D}$. Thus, the sample space of D is given by $\{\emptyset, \mathscr{D}_1, \mathscr{D}_2, \cdots, \mathscr{D}_n, \cdots, \mathscr{D}_{2^N-1}\}$, where $\emptyset$ denotes an empty set and $\mathscr{D}_n$ denotes the n-th non-empty subcollection of the N relay nodes. Specifically, the event $\mathscr{D} = \emptyset$ means that all the relays fail to decode x_s, which can be described as

$$C_{sr_i} < R, \quad i = 1, 2, \cdots, N, \tag{5.7}$$

where R is the data rate and C_{sr_i} is given by (5.2). Meanwhile, the event $\mathscr{D} = \mathscr{D}_n$ implies that the relays within $\mathscr{D}_n$ successfully decode the source signal and the remaining relays fail to decode. In an information-theoretic sense, the event $\mathscr{D} = \mathscr{D}_n$ can thus be described as

$$C_{sr_i} > R, \quad r_i \in \mathscr{D}_n$$
$$C_{sr_j} < R, \quad r_j \in \bar{\mathscr{D}}_n. \tag{5.8}$$

where $\bar{\mathscr{D}}_n = \mathscr{R} - \mathscr{D}_n$ denotes the complementary set of $\mathscr{D}_n$. If the decoding set $\mathscr{D}$ is empty, implying that all the relays fail to decode, then no relay is selected to forward the source signal. If the decoding set $\mathscr{D}$ is non-empty e.g. $\mathscr{D} = \mathscr{D}_n$, a relay node is chosen from the decoding set to retransmit the source signal x_s. Without loss of generality, we consider that a relay R_i is selected among the decoding set. It needs to be pointed out that in the pure relay selection scheme, only a relay node is employed to assist the S-D transmission without relying on a jammer. Thus, the transmit power of the selected relay is given by P_s in the pure relay selection scheme, which is different from the joint relay and jammer selection, where the total power P_s is shared by the selected relay and jammer. Considering that the selected relay retransmits the source signal x_s at a power of P_s, we can express the received signal at the D as

$$y_d^{\text{relay}} = h_{r_id}\sqrt{P_s}x_s + n_d, \tag{5.9}$$

where h_{r_id} represents the wireless fading of the R_i-D channel and n_d represents the AWGN received at R_i. Meanwhile, the E overhears the relay transmission and the corresponding received signal is written as

$$y_e^{\text{relay}} = h_{r_ie}\sqrt{P_s}x_s + n_e, \tag{5.10}$$

where h_{r_ie} represents the wireless fading of the R_i-E channel and n_e represents the AWGN received at the E. From (5.9), we obtain the capacity of the R_i-D channel as

$$C_{r_id}^{\text{relay}} = \frac{1}{2}\log_2(1 + |h_{r_id}|^2\gamma_s). \tag{5.11}$$

Similarly, from (5.10), we obtain the capacity of the R_i-E channel as

$$C_{r_ie}^{\text{relay}} = \frac{1}{2}\log_2(1 + |h_{r_ie}|^2\gamma_s). \tag{5.12}$$

By considering that the passive eavesdropper's CSI is unavailable, a relay node that maximizes $C_{r_id}^{\text{relay}}$ is regarded as the best "relay" to participate in forwarding the source transmission. Thus, from (5.11), the relay selection criterion is given by

$$r = \arg\max_{r_i\in\mathcal{D}_n} C_{r_id} = \arg\max_{r_i\in\mathcal{D}_n} |h_{r_id}|^2, \tag{5.13}$$

where r denotes the selected relay and $\mathcal{D}_n$ denotes the decoding set. Denoting the selected relay by 'r', we can obtain the channel capacity from the selected relay to D and that from the selected relay to E as

$$C_{rd} = \max_{r_i\in\mathcal{D}_n} \frac{1}{2}\log_2(1 + |h_{r_id}|^2\gamma_s), \tag{5.14}$$

and

$$C_{re} = \frac{1}{2}\log_2(1 + |h_{re}|^2\gamma_s), \tag{5.15}$$

where $|h_{re}|^2$ represents the wireless fading of the channel spanning from the selected relay to E. As defined in (4.8) and (4.9), when the channel capacity achieved at the D drops below the data rate R, an outage event is considered to occur, while an intercept event happens when the channel capacity achieved at the E becomes higher than the data rate. Therefore, the outage probability of the S-D transmission relying on the pure relay selection scheme can be obtained as

$$P_{\text{out}}^{\text{relay}} = \sum_{n=1}^{2^N-1} \Pr(C_{rd} < R, \mathscr{D} = \mathscr{D}_n), \tag{5.16}$$

where C_{rd} is given by (5.14). Meanwhile, the intercept probability of the conventional pure relay selection is given by

$$P_{\text{int}}^{\text{relay}} = \sum_{n=1}^{2^N-1} \Pr(C_{re} > R, \mathscr{D} = \mathscr{D}_n), \tag{5.17}$$

where C_{re} is given by (5.15). Substituting C_{rd} from (5.14) into (5.16) gives

$$P_{\text{out}}^{\text{relay}} = \sum_{n=1}^{2^N-1} \Pr(\mathscr{D} = \mathscr{D}_n) \Pr(\max_{r_i \in \mathscr{D}_n} |h_{r_id}|^2 < \Delta), \tag{5.18}$$

where $\Delta = (2^{2R} - 1)/\gamma_s$. Since the random variable $|h_{r_id}|^2$ is exponentially distributed and independent for different R_i, we have

$$P_{\text{out}}^{\text{relay}} = \sum_{n=1}^{2^N-1} \Pr(\mathscr{D} = \mathscr{D}_n) \prod_{r_i \in \mathscr{D}_n} [1 - \exp(-\frac{\Delta}{\sigma_{r_id}^2})], \tag{5.19}$$

where the term $\Pr(\mathscr{D} = \mathscr{D}_n)$ can be obtained from (5.8) as

$$\Pr(\mathscr{D} = \mathscr{D}_n) = \prod_{r_i \in \mathscr{D}_n} \Pr(C_{sr_i} > R) \prod_{r_j \in \bar{\mathscr{D}}_n} \Pr(C_{sr_j} < R). \tag{5.20}$$

Substituting C_{sr_i} from (5.2) into the term $\Pr(C_{sr_i} > R)$ yields

$$\Pr(C_{sr_i} > R) = \Pr(|h_{sr_i}|^2 > \Delta) = \exp(-\frac{\Delta}{\sigma_{sr_i}^2}), \tag{5.21}$$

where $\sigma_{sr_i}^2$ denotes the mean of random variable $|h_{sr_i}|^2$. Similarly, the term $\Pr(C_{sr_j} < R)$ can be given by

$$\Pr(C_{sr_j} < R) = 1 - \exp(-\frac{\Delta}{\sigma_{sr_j}^2}). \tag{5.22}$$

Substituting $\Pr(C_{sr_i} > R)$ and $\Pr(C_{sr_j} < R)$ from (5.21) and (5.22) into (5.20) gives

$$\Pr(\mathscr{D} = \mathscr{D}_n) = \prod_{r_i \in \mathscr{D}_n} \exp(-\frac{\Delta}{\sigma_{sr_i}^2}) \prod_{r_j \in \bar{\mathscr{D}}_n} [1 - \exp(-\frac{\Delta}{\sigma_{sr_j}^2})]. \tag{5.23}$$

Substituting $\Pr(\mathscr{D} = \mathscr{D}_n)$ from (5.23) into (5.19), we arrive at

$$P_{\text{out}}^{\text{relay}} = \sum_{n=1}^{2^N-1} \prod_{r_i \in \mathscr{D}_n} \exp(-\frac{\Delta}{\sigma_{sr_i}^2})[1 - \exp(-\frac{\Delta}{\sigma_{r_id}^2})] \prod_{r_j \in \bar{\mathscr{D}}_n} [1 - \exp(-\frac{\Delta}{\sigma_{sr_j}^2})], \qquad (5.24)$$

which gives a closed-form outage probability for the pure relay selection scheme. Additionally, from (5.17), the intercept probability of the pure relay selection scheme can be obtained as

$$P_{\text{int}}^{\text{relay}} = \sum_{n=1}^{2^N-1} \Pr(\mathscr{D} = \mathscr{D}_n) \sum_{r_i \in \mathscr{D}_n} \Pr(C_{r_ie} > R, r = r_i), \qquad (5.25)$$

where the term $\Pr(\mathscr{D} = \mathscr{D}_n)$ is given by (5.23). Moreover, by combining (5.13) and (5.15), the term $\Pr(C_{r_ie} > R, r = r_i)$ can be given by

$$\Pr(C_{r_ie} > R, r = i) = \Pr(|h_{r_ie}|^2 > \Delta, |h_{r_id}|^2 > \max_{r_j \in \mathscr{D}_n - \{r_i\}} |h_{r_jd}|^2), \qquad (5.26)$$

where $\Delta = (2^{2R} - 1)/\gamma_s$ and $\mathscr{D}_n - \{r_i\}$ represents the set difference between $\mathscr{D}_n$ and $\{r_i\}$. Considering that the exponential random variables $|h_{r_ie}|^2$ and $|h_{r_id}|^2$ are independent of each other, we can rewrite the preceding equation as

$$\Pr(C_{r_ie} > R, r = r_i) = \Pr(|h_{r_ie}|^2 > \Delta)\Pr(|h_{r_id}|^2 > \max_{r_j \in \mathscr{D}_n - \{r_i\}} |h_{r_jd}|^2), \qquad (5.27)$$

where the term $\Pr(|h_{r_ie}|^2 > \Delta)$ is given by

$$\Pr(|h_{r_ie}|^2 > \Delta) = \exp(-\frac{\Delta}{\sigma_{r_ie}^2}). \qquad (5.28)$$

Denoting $|h_{r_id}|^2 = X$ and $E(|h_{r_id}|^2) = \sigma_{r_id}^2$, we obtain the PDF of X as

$$p_X(x) = \frac{1}{\sigma_{r_id}^2} \exp(-\frac{x}{\sigma_{r_id}^2}), \qquad (5.29)$$

for $x > 0$. Noting that the random variable $|h_{r_jd}|^2$ is independent for different relays, we obtain the term $\Pr(|h_{r_id}|^2 > \max_{r_j \in \mathscr{D}_n - \{r_i\}} |h_{r_jd}|^2)$ as

$$\Pr(|h_{r_id}|^2 > \max_{r_j \in \mathscr{D}_n - \{r_i\}} |h_{r_jd}|^2) = \int_0^\infty \prod_{r_j \in \mathscr{D}_n - \{r_i\}} [1 - \exp(-\frac{x}{\sigma_{r_jd}^2})]p_X(x)dx, \qquad (5.30)$$

where $\sigma_{r_jd}^2$ is the mean of random variable $|h_{r_jd}|^2$. From the binomial expansion, the term $\prod_{r_j \in \mathcal{D}_n - \{r_i\}} [1 - \exp(-\frac{x}{\sigma_{r_jd}^2})]$ is given by

$$\prod_{r_j \in \mathcal{D}_n - \{r_i\}} [1 - \exp(-\frac{x}{\sigma_{r_jd}^2})] = 1 + \sum_{k=1}^{2^{|\mathcal{D}_n|-1}-1} (-1)^{|\mathcal{C}_{n,k}|} \exp(- \sum_{r_j \in \mathcal{C}_{n,k}} \frac{x}{\sigma_{r_jd}^2}), \qquad (5.31)$$

where $\mathcal{C}_{n,k}$ denotes the k-th non-empty collection of the set $\mathcal{D}_n - \{r_i\}$ and $|\cdot|$ denotes the set cardinality. Substituting $p_X(x)$ and $\prod_{r_j \in \mathcal{D}_n - \{r_i\}} [1 - \exp(-\frac{x}{\sigma_{r_jd}^2})]$ from (5.29) and (5.31) into (5.30) yields

$$\Pr(|h_{r_id}|^2 > \max_{r_j \in \mathcal{D}_n - \{r_i\}} |h_{r_jd}|^2) = 1$$

$$+ \sum_{k=1}^{2^{|\mathcal{D}_n|-1}-1} (-1)^{|\mathcal{C}_{n,k}|} \int_0^\infty \frac{1}{\sigma_{r_id}^2} \exp(-\frac{x}{\sigma_{r_id}^2} - \sum_{r_j \in \mathcal{C}_{n,k}} \frac{x}{\sigma_{r_jd}^2}) dx, \qquad (5.32)$$

which can be further obtained as

$$\Pr(|h_{r_id}|^2 > \max_{r_j \in \mathcal{D}_n - \{r_i\}} |h_{r_jd}|^2) = 1 + \sum_{k=1}^{2^{|\mathcal{D}_n|-1}-1} (-1)^{|\mathcal{C}_{n,k}|} (1 + \sum_{r_j \in \mathcal{C}_{n,k}} \frac{\sigma_{r_id}^2}{\sigma_{r_jd}^2})^{-1}. \qquad (5.33)$$

Combining (5.28) and (5.33) with (5.27), we arrive at

$$\Pr(C_{r_ie} > R, r = r_i) = \exp(-\frac{\Delta}{\sigma_{r_ie}^2})[1 + \sum_{k=1}^{2^{|\mathcal{D}_n|-1}-1} (-1)^{|\mathcal{C}_{n,k}|} (1 + \sum_{r_j \in \mathcal{C}_{n,k}} \frac{\sigma_{r_id}^2}{\sigma_{r_jd}^2})^{-1}]. \qquad (5.34)$$

Substituting $\Pr(\mathcal{D} = \mathcal{D}_n)$ and $\Pr(C_{r_ie} > R, r = r_i)$ from (5.23) and (5.34) into (5.25) gives

$$P_{\text{int}}^{\text{relay}} = \sum_{n=1}^{2^N-1} \prod_{r_i \in \mathcal{D}_n} \exp(-\frac{\Delta}{\sigma_{sr_i}^2}) \prod_{r_j \in \bar{\mathcal{D}}_n} [1 - \exp(-\frac{\Delta}{\sigma_{sr_j}^2})]$$

$$\times \sum_{r_i \in \mathcal{D}_n} \exp(-\frac{\Delta}{\sigma_{r_ie}^2})[1 + \sum_{k=1}^{2^{|\mathcal{D}_n|-1}-1} (-1)^{|\mathcal{C}_{n,k}|} (1 + \sum_{r_j \in \mathcal{C}_{n,k}} \frac{\sigma_{r_id}^2}{\sigma_{r_jd}^2})^{-1}], \qquad (5.35)$$

which is a closed-form intercept probability expression for the pure relay selection scheme.

5.2.2 Pure Jammer Selection

In this subsection, we present the pure jammer selection scheme, where the S directly transmits to the D without relying on the relay, while a friendly jammer is chosen among the M jammer candidates to transmit the artificial noise for protecting the S-D transmission against eavesdropping. In order to make a fair comparison with the pure relay selection in terms of power consumption, the total transmit power of the source and selected jammer is constrained to P_s. For simplicity, we consider the simple equal power allocation and the transmit powers of the source and jammer are given by $P_s/2$. Again, following [10–12], the selected jammer is assumed to transmit an artificial noise that only interferes with the E without affecting the legitimate D. Thus, considering that the S directly transmits x_s to the D at a power of $P_s/2$, we can express the received signal at the D as

$$y_d^{\text{jammer}} = h_{sd}\sqrt{\frac{P_s}{2}}x_s + n_d, \tag{5.36}$$

where h_{sd} represents the wireless fading of the S-D channel and n_d represents the AWGN received at D. Using (5.36), we can obtain the capacity of the S-D channel as

$$C_{sd}^{\text{jammer}} = \log_2(1 + |h_{sd}|^2\frac{\gamma_s}{2}). \tag{5.37}$$

Meanwhile, the E is assumed to be capable of overhearing the S-D transmission, which is however, corrupted by the selected jammer as denoted by J_m with the aid of transmitting the artificial noise at a power of $P_s/2$. Hence, the received signal at the E can be expressed as

$$y_e^{\text{jammer}} = h_{se}\sqrt{\frac{P_s}{2}}x_s + h_{j_me}\sqrt{\frac{P_s}{2}}x_n + n_e, \tag{5.38}$$

where h_{se} and h_{j_me} represent the wireless fading of the S-E channel and that of the J_m-E channel, x_n represents the artificial noise, and n_e represents the AWGN at E. From (5.38), the capacity of the S-E channel with the help of the selected jammer j_m as

$$C_{se}^{\text{jammer}}(j_m) = \log_2(1 + \frac{|h_{se}|^2\gamma_s}{|h_{j_me}|^2\gamma_s + 2}). \tag{5.39}$$

Here, we assume that the friendly jammers' CSI knowledge is available, which may be obtained through using a channel estimation method [13–15]. In generally, a friendly jammer that minimizes the channel capacity achieved at the E C_{se}^{jammer} is selected to transmit the artificial noise for interfering with the E. Hence, from (5.39), we obtain the jammer selection criterion as

$$j = \arg\min_{j_m \in \mathscr{J}} C_{se}^{jammer}(j_m) = \arg\max_{j_m \in \mathscr{J}} |h_{j_m e}|^2, \tag{5.40}$$

where j denotes the selected jammer and $\mathscr{J}$ denotes the set of M friendly jammers. Thus, combining (5.39) and (5.40), the capacity achieved by the E relying on the pure jammer selection is given by

$$C_{se}^{jammer} = \min_{j_m \in \mathscr{J}} \log_2(1 + \frac{|h_{se}|^2 \gamma_s}{|h_{j_m e}|^2 \gamma_s + 2}). \tag{5.41}$$

Again, following the outage definition as given by (4.8), when the channel capacity achieved at the D becomes less than the data rate R, an outage event happens. Hence, using (5.37), we obtain the outage probability of the pure jammer selection as

$$P_{out}^{jammer} = \Pr(C_{sd}^{jammer} < R), \tag{5.42}$$

which is further given by

$$P_{out}^{jammer} = \Pr(|h_{sd}|^2 < 2\Lambda), \tag{5.43}$$

where $\Lambda = (2^R - 1)/\gamma_s$. Noting that $|h_{sd}|^2$ is an exponentially distributed random variable with the mean of σ_{sd}^2, we arrive at

$$P_{out}^{jammer} = 1 - \exp(-\frac{2\Lambda}{\sigma_{sd}^2}), \tag{5.44}$$

which gives a closed-form outage probability expression for the pure jammer selection scheme. Additionally, according to the intercept definition of (4.9), the intercept probability of the pure jammer selection scheme can be obtained from (5.41) as

$$P_{int}^{jammer} = \Pr(C_{se}^{jammer} > R), \tag{5.45}$$

which can be rewritten as

$$P_{int}^{jammer} = \Pr(\max_{j_m \in \mathscr{J}} |h_{j_m e}|^2 \gamma_s \Lambda < |h_{se}|^2 - 2\Lambda). \tag{5.46}$$

where $\Lambda = (2^R - 1)/\gamma_s$. By denoting $|h_{se}|^2 = Y$, the PDF of Y is expressed as

$$p_Y(y) = \frac{1}{\sigma_{se}^2} \exp(-\frac{y}{\sigma_{se}^2}), \tag{5.47}$$

for $y > 0$. We can thus rewrite (5.46) as

$$P_{\text{int}}^{\text{jammer}} = \int_{2\Lambda}^{\infty} \Pr(\max_{j_m \in \mathscr{J}} |h_{j_m e}|^2 \gamma_s \Lambda < y - 2\Lambda) p_Y(y) dy, \qquad (5.48)$$

where $p_Y(y)$ is given by (5.47). Noting that $|h_{j_m e}|^2$ is an exponential random variable with the mean of $\sigma_{j_m e}^2$ and independent for different jammers, we have

$$P_{\text{int}}^{\text{jammer}} = \int_{2\Lambda}^{\infty} \prod_{j_m \in \mathscr{J}} [1 - \exp(-\frac{y - 2\Lambda}{\sigma_{j_m e}^2 \gamma_s \Lambda})] p_Y(y) dy, \qquad (5.49)$$

where the term $\prod_{j_m \in \mathscr{J}} [1 - \exp(-\frac{y-2\Lambda}{\sigma_{j_m e}^2 \gamma_s \Lambda})]$ can be expanded as

$$\prod_{j_m \in \mathscr{J}} [1 - \exp(-\frac{y - 2\Lambda}{\sigma_{j_m e}^2 \gamma_s \Lambda})] = 1 + \sum_{k=1}^{2^M - 1} (-1)^{|\mathscr{Z}_k|} \exp(- \sum_{j_m \in \mathscr{Z}_k} \frac{y - 2\Lambda}{\sigma_{j_m e}^2 \gamma_s \Lambda}), \qquad (5.50)$$

where $\mathscr{Z}_k$ denotes the k-th non-empty subcollection of the jammer set $\mathscr{J}$. Combining (5.49) and (5.50), we obtain

$$P_{\text{int}}^{\text{jammer}} = \int_{2\Lambda}^{\infty} p_Y(y) dy + \int_{2\Lambda}^{\infty} \sum_{k=1}^{2^M - 1} (-1)^{|\mathscr{Z}_k|} \exp(- \sum_{j_m \in \mathscr{Z}_k} \frac{y - 2\Lambda}{\sigma_{j_m e}^2 \gamma_s \Lambda}) p_Y(y) dy. \qquad (5.51)$$

Substituting $p_Y(y)$ from (5.47) into the preceding equation gives

$$P_{\text{int}}^{\text{jammer}} = \int_{2\Lambda}^{\infty} \frac{1}{\sigma_{se}^2} \exp(-\frac{y}{\sigma_{se}^2})$$

$$+ \int_{2\Lambda}^{\infty} \sum_{k=1}^{2^M - 1} (-1)^{|\mathscr{Z}_k|} \frac{1}{\sigma_{se}^2} \exp(- \sum_{j_m \in \mathscr{Z}_k} \frac{y - 2\Lambda}{\sigma_{j_m e}^2 \gamma_s \Lambda} - \frac{y}{\sigma_{se}^2}) dy, \qquad (5.52)$$

which is obtained as

$$P_{\text{int}}^{\text{jammer}} = \exp(-\frac{2\Lambda}{\sigma_{se}^2}) + \sum_{k=1}^{2^M - 1} \int_{2\Lambda}^{\infty} \frac{(-1)^{|\mathscr{Z}_k|}}{\sigma_{se}^2} \exp(- \sum_{j_m \in \mathscr{Z}_k} \frac{y - 2\Lambda}{\sigma_{j_m e}^2 \gamma_s \Lambda} - \frac{y}{\sigma_{se}^2}) dy. \qquad (5.53)$$

Performing the integration of (5.53) yields a closed-form intercept probability expression for the pure jammer selection as

$$P_{\text{int}}^{\text{jammer}} = \exp(-\frac{2\Lambda}{\sigma_{se}^2}) + \sum_{k=1}^{2^M - 1} (-1)^{|\mathscr{Z}_k|} (1 + \sum_{j_m \in \mathscr{Z}_k} \frac{\sigma_{se}^2}{\sigma_{j_m e}^2 \gamma_s \Lambda})^{-1} \exp(-\frac{2\Lambda}{\sigma_{se}^2}), \qquad (5.54)$$

which completes the SRT analysis of the conventional pure jammer selection scheme.

5.2.3 Joint Relay and Jammer Selection

This subsection presents the joint relay and jammer selection scheme for the sake of enhancing the SRT performance of the *S-D* transmission with the aid of N relay nodes and M friendly jammers, among which a relay is selected to assist the source transmission and meanwhile, a friendly jammer is chosen to transmit the artificial noise for interfering with the E. Generally speaking, a relay node that successfully decodes the source signal and maximizes the channel capacity C_{r_id} as given by (5.5) is considered to participate in forwarding the source signal to the D. Meanwhile, a friendly jammer of minimizing the E's channel capacity $C_{r_ie}(j_m)$ is chosen to emit the artificial noise. Thus, by using (5.5) and (5.6), the joint relay and jammer selection criterion can be expressed as

$$r = \arg\max_{r_i \in \mathscr{D}_n} C_{r_id} = \arg\max_{r_i \in \mathscr{D}_n} |h_{r_id}|^2,$$

$$j = \arg\min_{j_m \in \mathscr{J}} C_{r_ie}(j_m) = \arg\max_{j_m \in \mathscr{J}} |h_{j_me}|^2, \tag{5.55}$$

where $\mathscr{D}_n$ denotes the decoding set, $\mathscr{J}$ denotes the jammer set, r denotes the selected relay, and j denotes the selected jammer. Combining (5.5), (5.6) and (5.55), we can obtain the capacity achieved at the D and E as

$$C_{rd}^{\text{joint}} = \max_{r_i \in \mathscr{D}_n} \frac{1}{2}\log_2(1 + |h_{r_id}|^2\frac{\gamma_s}{2}), \tag{5.56}$$

and

$$C_{re}^{\text{joint}} = \min_{j_m \in \mathscr{J}} \frac{1}{2}\log_2(1 + \frac{|h_{re}|^2\gamma_s}{|h_{j_me}|^2\gamma_s + 2}), \tag{5.57}$$

where h_{re} denotes the fading of the channel spanning from the selected relay r to E. Similarly to (5.17), the outage probability of the proposed joint relay and jammer selection is obtained as

$$P_{\text{out}}^{\text{joint}} = \sum_{n=1}^{2^N-1} \Pr(\mathscr{D} = \mathscr{D}_n) \Pr(C_{rd}^{\text{joint}} < R), \tag{5.58}$$

where C_{rd}^{joint} is given by (5.56). Also, the intercept probability of the proposed joint relay and jammer selection can be expressed as

$$P_{\text{int}}^{\text{joint}} = \sum_{n=1}^{2^N-1} \Pr(\mathscr{D} = \mathscr{D}_n) \Pr(C_{re}^{\text{joint}} > R), \tag{5.59}$$

where C_{re}^{joint} is given by (5.57). Substituting C_{rd}^{joint} from (5.56) into (5.58) gives

$$
P_{\text{out}}^{\text{joint}} = \sum_{n=1}^{2^N-1} \Pr(\mathscr{D} = \mathscr{D}_n) \Pr\left[\max_{r_i \in \mathscr{D}_n} \frac{1}{2}\log_2(1 + |h_{r_id}|^2\frac{\gamma_s}{2}) < R\right], \tag{5.60}
$$

which is rewritten as

$$
P_{\text{out}}^{\text{joint}} = \sum_{n=1}^{2^N-1} \Pr(\mathscr{D} = \mathscr{D}_n) \prod_{r_i \in \mathscr{D}_n} \Pr(|h_{r_id}|^2 < 2\Delta), \tag{5.61}
$$

where $\Delta = (2^{2R} - 1)/\gamma_s$. Noting that $|h_{r_id}|^2$ is an exponentially distributed random variable with a mean of $\sigma_{r_id}^2$, we obtain

$$
\Pr(|h_{r_id}|^2 < 2\Delta) = 1 - \exp(-\frac{2\Delta}{\sigma_{r_id}^2}). \tag{5.62}
$$

Thus, substituting $\Pr(\mathscr{D} = \mathscr{D}_n)$ and $\Pr(|h_{r_id}|^2 < 2\Delta)$ from (5.23) and (5.62) into (5.61) yields

$$
P_{\text{out}}^{\text{joint}} = \sum_{n=1}^{2^N-1} \prod_{r_i \in \mathscr{D}_n} \exp(-\frac{\Delta}{\sigma_{sr_i}^2})[1 - \exp(-\frac{2\Delta}{\sigma_{r_id}^2})] \prod_{r_j \in \bar{\mathscr{D}}_n} [1 - \exp(-\frac{\Delta}{\sigma_{sr_j}^2})], \tag{5.63}
$$

which gives a closed-form expression of the outage probability for proposed joint relay and jammer selection scheme. Additionally, substituting C_{re}^{joint} from (5.57) into (5.59), we have

$$
P_{\text{int}}^{\text{joint}} = \sum_{n=1}^{2^N-1} \Pr(\mathscr{D} = \mathscr{D}_n) \Pr\left[\min_{j_m \in \mathscr{J}} \frac{1}{2}\log_2(1 + \frac{|h_{re}|^2\gamma_s}{|h_{j_me}|^2\gamma_s + 2}) > R\right], \tag{5.64}
$$

which can be rewritten as

$$
P_{\text{int}}^{\text{joint}} = \sum_{n=1}^{2^N-1} \Pr(\mathscr{D} = \mathscr{D}_n) \Pr(\max_{j_m \in \mathscr{J}} |h_{j_me}|^2 < \frac{|h_{re}|^2}{\Delta\gamma_s} - \frac{2}{\gamma_s}), \tag{5.65}
$$

where $\Delta = (2^{2R} - 1)/\gamma_s$ and $\Pr(\mathscr{D} = \mathscr{D}_n)$ is given by (5.23). By using the law of total probability, the term $\Pr(\max_{j_m \in \mathscr{J}} |h_{j_me}|^2 < \frac{|h_{re}|^2}{\Delta\gamma_s} - \frac{2}{\gamma_s})$ is obtained as

$$
\Pr(\max_{j_m \in \mathscr{J}} |h_{j_me}|^2 < \frac{|h_{re}|^2}{\Delta\gamma_s} - \frac{2}{\gamma_s}) = \sum_{r_i \in \mathscr{D}_n} \Pr(\max_{j_m \in \mathscr{J}} |h_{j_me}|^2 < \frac{|h_{r_ie}|^2}{\Delta\gamma_s} - \frac{2}{\gamma_s}, r = r_i),
$$
$$\tag{5.66}$$

where r denotes the selected relay. Combining (5.55) and (5.66), we have

$$\Pr(\max_{jm \in \mathscr{J}} |h_{jme}|^2 < \frac{|h_{re}|^2}{\Delta \gamma_s} - \frac{2}{\gamma_s}) =$$

$$\sum_{r_i \in \mathscr{D}_n} \Pr(\max_{jm \in \mathscr{J}} |h_{jme}|^2 < \frac{|h_{r_ie}|^2}{\Delta \gamma_s} - \frac{2}{\gamma_s}, |h_{r_id}|^2 > \max_{r_j \in \mathscr{D}_n - \{r_i\}} |h_{r_jd}|^2), \quad (5.67)$$

where $\mathscr{D}_n - \{r_i\}$ denotes the set difference between $\mathscr{D}_n$ and $\{r_i\}$. Since random variables $|h_{r_ie}|^2$, $|h_{r_id}|^2$, $|h_{r_jd}|^2$ and $|h_{jme}|^2$ are independent of each other, we can rewrite (5.67) as

$$\Pr(\max_{jm \in \mathscr{J}} |h_{jme}|^2 < \frac{|h_{re}|^2}{\Delta \gamma_s} - \frac{2}{\gamma_s}) = \sum_{r_i \in \mathscr{D}_n} \Pr(\max_{jm \in \mathscr{J}} |h_{jme}|^2 < \frac{|h_{r_ie}|^2}{\Delta \gamma_s} - \frac{2}{\gamma_s})$$

$$\times \Pr(|h_{r_id}|^2 > \max_{r_j \in \mathscr{D}_n - \{r_i\}} |h_{r_jd}|^2). \quad (5.68)$$

Denoting $|h_{r_ie}|^2 = Z$ and $\sigma_{r_ie}^2 = E(|h_{r_ie}|^2)$, the PDF of Z can be expressed as

$$p_Z(z) = \frac{1}{\sigma_{r_ie}^2} \exp(-\frac{z}{\sigma_{r_ie}^2}), \quad (5.69)$$

for $z > 0$. Thus, the term $\Pr(\max_{jm \in \mathscr{J}} |h_{jme}|^2 < \frac{|h_{r_ie}|^2}{\Delta \gamma_s} - \frac{2}{\gamma_s})$ can be obtained as

$$\Pr(\max_{jm \in \mathscr{J}} |h_{jme}|^2 < \frac{|h_{r_ie}|^2}{\Delta \gamma_s} - \frac{2}{\gamma_s}) = \int_{2\Delta}^{\infty} \Pr(\max_{jm \in \mathscr{J}} |h_{jme}|^2 < \frac{z}{\Delta \gamma_s} - \frac{2}{\gamma_s}) p_Z(z) dz,$$

$$(5.70)$$

which is further given by

$$\Pr(\max_{jm \in \mathscr{J}} |h_{jme}|^2 < \frac{|h_{r_ie}|^2}{\Delta \gamma_s} - \frac{2}{\gamma_s}) = \int_{2\Delta}^{\infty} \prod_{m=1}^{M} [1 - \exp(-\frac{z}{\Delta \gamma_s \sigma_{jme}^2} + \frac{2}{\gamma_s \sigma_{jme}^2})] p_Z(z) dz.$$

$$(5.71)$$

Using the binomial expansion, we can obtain the term $\prod_{m=1}^{M} [1 - \exp(-\frac{z}{\Delta \gamma_s \sigma_{jme}^2} + \frac{2}{\gamma_s \sigma_{jme}^2})]$ as

$$\prod_{m=1}^{M} [1 - \exp(-\frac{z}{\Delta \gamma_s \sigma_{jme}^2} + \frac{2}{\gamma_s \sigma_{jme}^2})] =$$

$$1 + \sum_{k=1}^{2^M - 1} (-1)^{|\mathscr{Z}_k|} \exp(-\sum_{jm \in \mathscr{Z}_k} \frac{z}{\Delta \gamma_s \sigma_{jme}^2} + \sum_{jm \in \mathscr{Z}_k} \frac{2}{\gamma_s \sigma_{jme}^2}), \quad (5.72)$$

where $\mathscr{Z}_k$ denotes the k-th non-empty subcollection of the M friendly jammers. Substituting $\prod_{m=1}^{M} [1 - \exp(-\frac{z}{\Delta\gamma_s\sigma_{jme}^2} + \frac{2}{\gamma_s\sigma_{jme}^2})]$ from (5.72) into (5.71) gives

$$\Pr(\max_{jm\in\mathscr{J}} |h_{jme}|^2 < \frac{|h_{rie}|^2}{\Delta\gamma_s} - \frac{2}{\gamma_s}) = \int_{2\Delta}^{\infty} p_Z(z)dz$$

$$+ \int_{2\Delta}^{\infty} \sum_{k=1}^{2^M-1} (-1)^{|\mathscr{Z}_k|} \exp(-\sum_{jm\in\mathscr{Z}_k} \frac{z}{\Delta\gamma_s\sigma_{jme}^2} + \sum_{jm\in\mathscr{Z}_k} \frac{2}{\gamma_s\sigma_{jme}^2})p_Z(z)dz. \quad (5.73)$$

Combining (5.69) and (5.73), we have

$$\Pr(\max_{jm\in\mathscr{J}} |h_{jme}|^2 < \frac{|h_{rie}|^2}{\Delta\gamma_s} - \frac{2}{\gamma_s}) = \int_{2\Delta}^{\infty} \frac{1}{\sigma_{rie}^2} \exp(-\frac{z}{\sigma_{rie}^2})dz$$

$$+ \sum_{k=1}^{2^M-1} \frac{(-1)^{|\mathscr{Z}_k|}}{\sigma_{rie}^2} \exp(\sum_{jm\in\mathscr{Z}_k} \frac{2}{\gamma_s\sigma_{jme}^2}) \int_{2\Delta}^{\infty} \exp(-\frac{z}{\sigma_{rie}^2} - \sum_{jm\in\mathscr{Z}_k} \frac{z}{\Delta\gamma_s\sigma_{jme}^2})dz, \quad (5.74)$$

which is obtained as

$$\Pr(\max_{jm\in\mathscr{J}} |h_{jme}|^2 < \frac{|h_{rie}|^2}{\Delta\gamma_s} - \frac{2}{\gamma_s}) = \exp(-\frac{2\Delta}{\sigma_{rie}^2})$$

$$+ \sum_{k=1}^{2^M-1} (-1)^{|\mathscr{Z}_k|}(1 + \sum_{jm\in\mathscr{Z}_k} \frac{\sigma_{rie}^2}{\Delta\gamma_s\sigma_{jme}^2})^{-1} \exp(-\frac{2\Delta}{\sigma_{rie}^2}). \quad (5.75)$$

Moreover, denoting $|h_{rid}|^2 = X$ and $\sigma_{rid}^2 = E(|h_{rid}|^2)$, the PDF of X is given by

$$p_X(x) = \frac{1}{\sigma_{rid}^2} \exp(-\frac{x}{\sigma_{rid}^2}), \quad (5.76)$$

for $x > 0$. We can thus rewrite the term $\Pr(|h_{rid}|^2 > \max_{r_j\in\mathscr{D}_n-\{r_i\}} |h_{rjd}|^2)$ as

$$\Pr(|h_{rid}|^2 > \max_{r_j\in\mathscr{D}_n-\{r_i\}} |h_{rjd}|^2) = \int_0^{\infty} \Pr(\max_{r_j\in\mathscr{D}_n-\{r_i\}} |h_{rjd}|^2 < x)p_X(x)dx, \quad (5.77)$$

which is given by

$$\Pr(|h_{rid}|^2 > \max_{r_j\in\mathscr{D}_n-\{r_i\}} |h_{rjd}|^2) = \int_0^{\infty} \prod_{r_j\in\mathscr{D}_n-\{r_i\}} [1 - \exp(-\frac{x}{\sigma_{rjd}^2})]p_X(x)dx. \quad (5.78)$$

Again, using the binomial expansion, the term $\prod_{r_j \in \mathcal{D}_n - \{r_i\}} [1 - \exp(-\frac{x}{\sigma_{r_j d}^2})]$ is obtained as

$$\prod_{r_j \in \mathcal{D}_n - \{r_i\}} [1 - \exp(-\frac{x}{\sigma_{r_j d}^2})] = 1 + \sum_{k=1}^{2^{|\mathcal{D}_n|-1}-1} (-1)^{|\mathcal{C}_{n,k}|} \exp(-\sum_{r_j \in \mathcal{C}_{n,k}} \frac{x}{\sigma_{r_j d}^2}), \quad (5.79)$$

where $\mathcal{C}_{n,k}$ is the k-th non-empty subcollection of the set $\mathcal{D}_n - \{r_i\}$. Substituting $\prod_{r_j \in \mathcal{D}_n - \{r_i\}} [1 - \exp(-\frac{x}{\sigma_{r_j d}^2})]$ from (5.79) into (5.78) gives

$$\Pr(|h_{r_i d}|^2 > \max_{r_j \in \mathcal{D}_n - \{r_i\}} |h_{r_j d}|^2) =$$

$$\int_0^\infty [1 + \sum_{k=1}^{2^{|\mathcal{D}_n|-1}-1} (-1)^{|\mathcal{C}_{n,k}|} \exp(-\sum_{r_j \in \mathcal{C}_{n,k}} \frac{x}{\sigma_{r_j d}^2})] p_X(x) dx, \quad (5.80)$$

which is rewritten as

$$\Pr(|h_{r_i d}|^2 > \max_{r_j \in \mathcal{D}_n - \{r_i\}} |h_{r_j d}|^2) = \int_0^\infty p_X(x) dx +$$

$$\int_0^\infty \sum_{k=1}^{2^{|\mathcal{D}_n|-1}-1} (-1)^{|\mathcal{C}_{n,k}|} \exp(-\sum_{r_j \in \mathcal{C}_{n,k}} \frac{x}{\sigma_{r_j d}^2}) p_X(x) dx. \quad (5.81)$$

Combining (5.76) and (5.81) yields,

$$\Pr(|h_{r_i d}|^2 > \max_{r_j \in \mathcal{D}_n - \{r_i\}} |h_{r_j d}|^2) = 1 +$$

$$\sum_{k=1}^{2^{|\mathcal{D}_n|-1}-1} \frac{(-1)^{|\mathcal{C}_{n,k}|}}{\sigma_{r_i d}^2} \int_0^\infty \exp(-\sum_{r_j \in \mathcal{C}_{n,k}} \frac{x}{\sigma_{r_j d}^2} - \frac{x}{\sigma_{r_i d}^2}) dx, \quad (5.82)$$

which is obtained as

$$\Pr(|h_{r_i d}|^2 > \max_{r_j \in \mathcal{D}_n - \{r_i\}} |h_{r_j d}|^2) = 1 + \sum_{k=1}^{2^{|\mathcal{D}_n|-1}-1} (-1)^{|\mathcal{C}_{n,k}|} (1 + \sum_{r_j \in \mathcal{C}_{n,k}} \frac{\sigma_{r_i d}^2}{\sigma_{r_j d}^2})^{-1}. \quad (5.83)$$

Substituting $\Pr(\max_{j_m \in \mathcal{J}} |h_{j_m e}|^2 < \frac{|h_{r_i e}|^2}{\Delta \gamma_s} - \frac{2}{\gamma_s})$ and $\Pr(|h_{r_i d}|^2 > \max_{r_j \in \mathcal{D}_n - \{r_i\}} |h_{r_j d}|^2)$ from (5.75) and (5.83) into (5.68) gives

$$\Pr(\max_{j_m \in \mathscr{J}} |h_{j_m e}|^2 < \frac{|h_{re}|^2}{\Delta \gamma_s} - \frac{2}{\gamma_s}) = \sum_{r_i \in \mathscr{D}_n} [1 + \sum_{k=1}^{2^{|\mathscr{D}_n|-1}-1} (-1)^{|\mathscr{C}_{n,k}|} (1 + \sum_{r_j \in \mathscr{C}_{n,k}} \frac{\sigma_{r_i d}^2}{\sigma_{r_j d}^2})^{-1}]$$

$$\times [\exp(-\frac{2\Delta}{\sigma_{r_i e}^2}) + \sum_{k=1}^{2^M-1} (-1)^{|\mathscr{Z}_k|} (1 + \sum_{j_m \in \mathscr{Z}_k} \frac{\sigma_{r_i e}^2}{\Delta \gamma_s \sigma_{j_m e}^2})^{-1} \exp(-\frac{2\Delta}{\sigma_{r_i e}^2})]$$

(5.84)

Finally, combining (5.23), (5.65) and (5.84), we arrive at

$$P_{\text{int}}^{\text{joint}} = \sum_{n=1}^{2^N-1} \prod_{r_i \in \mathscr{D}_n} \exp(-\frac{\Delta}{\sigma_{sr_i}^2}) \prod_{r_j \in \bar{\mathscr{D}}_n} [1 - \exp(-\frac{\Delta}{\sigma_{sr_j}^2})]$$

$$\times \sum_{r_i \in \mathscr{D}_n} [1 + \sum_{k=1}^{2^{|\mathscr{D}_n|-1}-1} (-1)^{|\mathscr{C}_{n,k}|} (1 + \sum_{r_j \in \mathscr{C}_{n,k}} \frac{\sigma_{r_i d}^2}{\sigma_{r_j d}^2})^{-1}]$$

$$\times [\exp(-\frac{2\Delta}{\sigma_{r_i e}^2}) + \sum_{k=1}^{2^M-1} (-1)^{|\mathscr{Z}_k|} (1 + \sum_{j_m \in \mathscr{Z}_k} \frac{\sigma_{r_i e}^2}{\Delta \gamma_s \sigma_{j_m e}^2})^{-1} \exp(-\frac{2\Delta}{\sigma_{r_i e}^2})], \quad (5.85)$$

which is a closed-form intercept probability expression for the proposed joint relay and jammer selection scheme. So far, we have derived closed-form expressions of the outage probability and intercept probability for the conventional pure relay selection and pure jammer selection as well as the proposed joint relay and jammer selection schemes over Rayleigh fading channels.

5.3 Numerical Results and Discussions

In this section, we present the numerical SRT results for the conventional direct transmission, pure relay selection, pure jammer selection as well as the proposed joint relay and jammer selection schemes. To be specific, the outage and intercept probabilities of the direct transmission, pure relay selection, pure jammer selection as well as the proposed joint relay and jammer selection are evaluated by using (4.17), (4.20), (5.24), (5.35), (5.44), (5.54), (5.63) and (5.85), respectively. In the numerical evaluation, the average gains of the main channel and wiretap channels are given by $\sigma_{sd}^2 = \sigma_{sr_i}^2 = \sigma_{sr_j}^2 = \sigma_{r_i d}^2 = \sigma_{r_j d}^2 = 1$ and $\sigma_{se}^2 = \sigma_{r_i e}^2 = \sigma_{j_m e}^2 = 0.1$. In addition, an SNR of $\gamma_s = 10 \, \text{dB}$, a data rate of $R = 0.5 \, \text{bit/s/Hz}$, the number of relays $N = 2$ and the number of jammers $M = 2$ are used, unless otherwise stated.

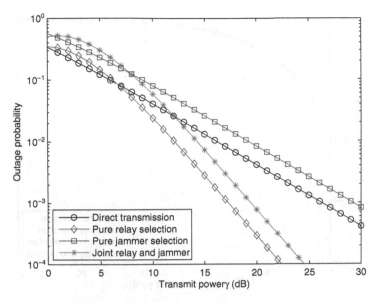

Fig. 5.2 Outage probability versus the SNR γ_s of the direct transmission, pure relay selection, pure jammer selection as well as the proposed joint relay and jammer selection schemes

Figure 5.2 shows the outage probability versus the SNR γ_s of the direct transmission, pure relay selection, pure jammer selection as well as the proposed joint relay and jammer selection schemes with $N = 2$ and $M = 2$, where N and M represent the number of relays and jammers, respectively. It can be seen from Fig. 5.2 that with an increasing SNR γ_s, the outage probabilities of the direct transmission, pure relay selection, pure jammer selection as well as the proposed joint relay and jammer selection schemes are reduced. This means that the reliability of wireless communications can be improved by increasing the transmit power. Figure 5.2 also shows that the outage performance of the pure jammer selection scheme is the worst and moreover, the proposed joint relay and jammer selection even performs worse than the direct transmission in terms of the outage probability in the low SNR region. This is because that in both the pure jammer selection as well as the joint relay and jammer selection schemes, a certain transit power is allocated to the selected jammer for sending the artificial noise, which is beneficial for the wireless security, but at the cost of the reliability degradation. Additionally, as shown in Fig. 5.2, the outage probability of the pure relay selection is higher than that of the direct transmission in the low SNR region, which arises from the fact that the half-duplex relaying constraint is considered.

Figure 5.3 depicts the intercept probability versus the SNR γ_s of the direct transmission, pure relay selection, pure jammer selection as well as the proposed joint relay and jammer selection schemes with $N = 2$ and $M = 2$. As shown in Fig. 5.3, as the SNR γ_s increases, the intercept probabilities of the direct transmission, pure relay selection, pure jammer selection as well as the proposed joint relay

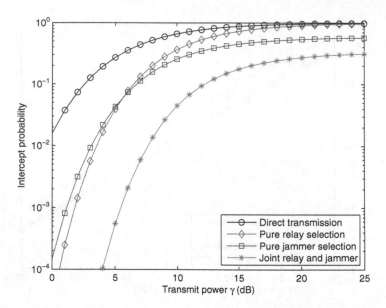

Fig. 5.3 Intercept probability versus the SNR γ_s of the direct transmission, pure relay selection, pure jammer selection as well as the proposed joint relay and jammer selection schemes

and jammer selection schemes increase accordingly. This means that increasing the transmit power leads to an increased intercept probability and thus degrades the wireless physical-layer security. Figure 5.3 also shows that the proposed joint relay and jammer selection outperforms the other schemes and the conventional direct transmission performs the worst in terms of their intercept probabilities, showing the security benefits of exploiting the relay and jammer selection. Moreover, one can see from Fig. 5.3 that as the SNR increases, the intercept performance of the pure jammer selection is initially worse and finally becomes better than that of the pure relay selection.

Figure 5.4 illustrates the outage probability versus the data rate R of the direct transmission, pure relay selection, pure jammer selection as well as the proposed joint relay and jammer selection schemes with $M = 2$ and $N = 2$. It can be seen from Fig. 5.4 that as the data rate R increases, the outage performance of the direct transmission, pure relay selection, pure jammer selection as well as the proposed joint relay and jammer selection schemes degrades accordingly. Figure 5.4 also shows that the pure relay selection as well as the proposed joint relay and jammer selection schemes outperform the direct transmission and pure jammer selection in terms of their outage probabilities in the low data rate region. Moreover, as the data rate R increases, the outage performance of the pure relay selection as well as the proposed joint relay and jammer selection schemes may become worse than that of the direct transmission and pure jammer selection. This is due to the fact that in both the pure relay selection as well as the proposed joint relay and jammer selection schemes, the half-duplex relaying constraint is considered, which

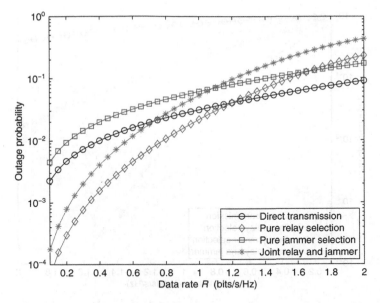

Fig. 5.4 Outage probability versus the data rate R of the direct transmission, pure relay selection, pure jammer selection as well as the proposed joint relay and jammer selection schemes

sacrifices the spectrum efficiency as compared to the direct transmission and pure jammer selection. Additionally, as shown in Fig. 5.4, the outage performance of the proposed joint relay and jammer selection is always worse than that of the pure relay selection, which is because that a certain transmit power is allocated to the selected friendly jammer for the physical-layer security improvement at the expense of the reliability degradation.

Figure 5.5 depicts the intercept probability versus the data rate R of the direct transmission, pure relay selection, pure jammer selection as well as the proposed joint relay and jammer selection schemes with $M = 2$ and $N = 2$. As shown in Fig. 5.5, as the data rate R increases, the intercept probabilities of the direct transmission, pure relay selection, pure jammer selection as well as the proposed joint relay and jammer selection schemes decrease. This is because that when an increased data rate was used by the source node, it becomes more difficult for an eavesdropper to successfully decode the source signal. It can also be seen from Fig. 5.5 that the intercept performance of the proposed joint relay and jammer selection is better than that of the other schemes, further confirming the security advantage of using the relay and jammer selection for defending against the eavesdropper.

Figure 5.6 shows the intercept probability versus outage probability of the direct transmission, pure relay selection, pure jammer selection as well as the proposed joint relay and jammer selection schemes with $N = 2$ and $M = 2$. One can observe from Fig. 5.6 that with an increasing outage probability, the intercept probabilities of the direct transmission, pure relay selection, pure jammer selection as well as the

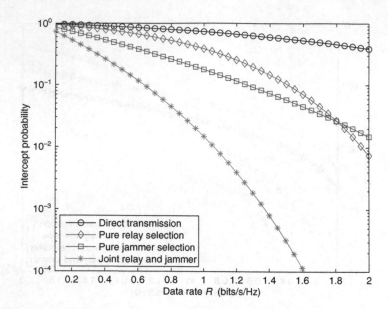

Fig. 5.5 Intercept probability versus the data rate R of the direct transmission, pure relay selection, pure jammer selection as well as the proposed joint relay and jammer selection schemes

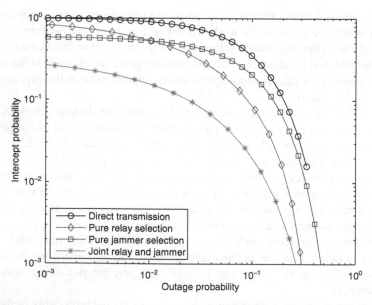

Fig. 5.6 Intercept probability versus outage probability of the direct transmission, pure relay selection, pure jammer selection as well as the proposed joint relay and jammer selection schemes

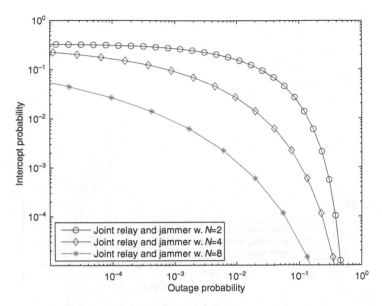

Fig. 5.7 Intercept probability versus outage probability of the joint relay and jammer selection for different N, where N represents the number of relay nodes

proposed joint relay and jammer selection schemes decrease accordingly and vice versa, showing the tradeoff between the security and reliability. Also, it is shown from Fig. 5.6 that the proposed joint relay and jammer selection scheme performs the best, while the conventional direct transmission is the worst in terms of their SRT performance.

In Fig. 5.7, we illustrate the intercept probability versus the outage probability of the joint relay and jammer selection scheme for different N, where N represents the number of relays. It can be observed from Fig. 5.7 that as the number of relays N increases from $N = 2$ to 8, the SRT of the proposed joint relay and jammer selection scheme improves significantly, confirming the security and reliability benefits of exploiting relay selection. In other words, increasing the number of relay nodes can effectively improve the wireless security and reliability simultaneously.

Figure 5.8 shows the intercept probability versus the outage probability of the joint relay and jammer selection scheme for different M, where M represents the number of friendly jammers. One can see from Fig. 5.8 that as the number of friendly jammers M increases from $M = 2$ to 8, the outage probability of the proposed joint relay and jammer selection decreases under a required intercept probability constraint. Conversely, given an outage probability requirement, the intercept probability of the joint relay and jammer selection scheme decreases, as the number of friendly jammers M increases from $M = 2$ to 8. As a consequence, both the security and reliability of wireless communications can be enhanced by increasing the number of friendly jammers.

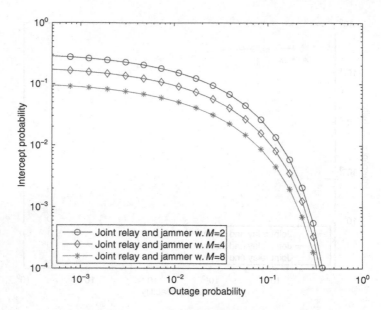

Fig. 5.8 Intercept probability versus outage probability of the joint relay and jammer selection for different M, where M represents the number of friendly jammers

5.4 Conclusions

In this chapter, we have studied the employment of joint relay and jammer selection for enhancing the security and reliability of a wireless network comprised of a source transmitting to a destination in the face of an eavesdropper, where multiple relays and jammers are utilized to assist the source-destination transmission against the eavesdropper. More specifically, a relay node is selected to help the source transmit to the destination, while a friendly jammer is used to generate the artificial noise for protecting the source transmission against eavesdropping. For the purpose of comparison, we have also considered the pure relay selection and pure jammer selection as benchmark schemes. We have carried out the SRT analysis for the proposed joint relay and jammer selection as well as the conventional pure relay selection and pure jammer selection over Rayleigh fading channels. It has been shown that the proposed joint relay and jammer selection outperforms the conventional pure relay selection and pure jammer selection in terms of their SRTs. In addition, with an increasing number of relay nodes and friendly jammers, the SRT performance of the proposed joint relay and jammer selection is improved significantly, showing the SRT enhancement of exploiting the relay and jammer selection.

References

1. Y. Zou, X. Wang, W. Shen, and L. Hanzo, "Security versus reliability analysis of opportunistic relaying," *IEEE Trans. Veh. Tech.*, vol. 63, no. 6, pp. 2653–2661, Jul. 2014.
2. J. Zhu, "Security-reliability trade-off for cognitive radio networks in the presence of eavesdropping attack," *EURASIP J. Adv. Signal Process.*, Jan. 2013
3. J. Zhu, Z. Liu, Y. Jiang, and Y. Zou, "Security-reliability tradeoff for relay selection in cooperative amplify-and-forward relay networks," in *Proc. 2015 Intern. Conf. Wirel. Commun. & Signal Process. (WCSP 2015)*, Nanjing China, Oct. 2015.
4. A. Bletsas, A. Khisti, D. P. Reed, and A. Lippman, "A simple cooperative diversity method based on network path selection," *IEEE J. Sel. Areas Commun.*, vol. 24, no. 3, pp. 659–672, Mar. 2006.
5. I. Krikidis, *et al.*, "Amplify-and-forward with partial relay selection," *IEEE Comm. Lett.*, vol. 12, no. 4, pp. 235–237, Apr. 2008.
6. Z. Yi and I.-M. Kim, "Diversity order analysis of the decode-and-forward cooperative networks with relay selection," *IEEE Trans. Wirel. Commun.*, vol. 7, no. 5, pp. 1792–1799, May 2008.
7. J. Vicario, *et al.*, "Opportunistic relay selection with outdated CSI: outage probability and diversity analysis," *IEEE Trans. Wirel. Commun.*, vol. 8, no. 6, pp. 2872–2876, Jun. 2009.
8. Y. Zou, J. Zhu, B. Zheng, and Y.-D. Yao, "An adaptive cooperation diversity scheme with best-relay selection in cognitive radio networks," *IEEE Trans. Signal Process.*, vol. 58, no. 10, pp. 5438–5445, Oct. 2010
9. Y. Ge, S. Wen, Y. H. Ang, and Y. C. Liang, "Optimal relay selection in IEEE 802.16 j multihop relay vehicular networks," *IEEE Trans. Veh. Tech.*, vol. 59, no. 5, pp. 2198–2206, May 2010.
10. S. Goel and R. Negi, "Guaranteeing secrecy using artificial noise," *IEEE Trans. Wirel. Commun.*, vol. 7, no. 6, pp. 2180–2189, Jul. 2008.
11. D. Goeckel, *et al.*, "Artificial noise generation from cooperative relays for everlasting secrecy in two-hop wireless networks," *IEEE J Sel. Areas Commun.*, vol. 29, no. 10, pp. 2067–2076, Oct. 2011.
12. N. Romero-Zurita, M. Ghogho, and D. McLernon, "Outage probability based power distribution between data and artificial noise for physical layer security," *IEEE Signal Process. Lett.*, vol. 19, no. 2, pp. 71–74, Feb. 2012.
13. O. Edfors, *et al.*, "OFDM channel estimation by singular value decomposition," *IEEE Trans. Commun.*, vol. 46, no. 7, pp. 931–939, Jul. 1998.
14. Y. Li, L. J. Cimini, and N. R. Sollenberger, "Robust channel estimation for OFDM systems with rapid dispersive fading channels," *IEEE Trans. Commun.*, vol. 46, no. 7, pp. 902–915, Jul. 1998.
15. M. Morelli and U. Mengali, "A comparison of pilot-aided channel estimation methods for OFDM systems," *IEEE Trans. Signal Process.*, vol. 49, no. 12, pp. 3065–3073, Dec. 2001.

Chapter 6
Summary

Abstract This chapter presents a summary about wireless physical-layer security for cooperative relay networks. We discuss the challenges and solutions of physical-layer security and their applications for cooperative relay networks. Specifically, the relay selection is studied for enhancing the physical-layer security of wireless communications in the presence of an eavesdropper. Also, a joint relay and jammer selection scheme is presented for wireless physical-layer security, where a relay is used to assist the transmission from the source to destination and meanwhile, a jammer is employed to transmit an artificial noise for confusing the eavesdropper. Additionally, the security-reliability tradeoff (SRT) is analytically formulated for wireless communications, where the security and reliability are measured by using the intercept probability encountered at the eavesdropper and the outage probability experienced at the legitimate destination, respectively. It is demonstrated that as the intercept probability increases, the outage probability decreases accordingly and vice versa, showing a tradeoff between the security and reliability. Moreover, two relay selection mechanisms, namely the singe-relay selection and multi-relay selection are devised for enhancing the wireless SRT performance. Finally, the joint relay and jammer selection is re-examined from the SRT perspective, which is capable of improving the wireless security and reliability concurrently.

6.1 Concluding Remarks

Due to the broadcast nature of radio propagation, wireless transmissions are open and accessible to any eavesdroppers. This makes the wireless communications become extremely vulnerable to the eavesdropping attacks. Traditionally, cryptographic techniques relying on secret keys have been adopted for protecting the confidentiality of wireless transmissions. There are two main types of cryptographic techniques, namely the public-key cryptography and symmetric-key cryptography, which are however only computationally secure and rely upon the hardness of their underlying mathematical problems. To this end, physical-layer security emerges as a promising paradigm of securing the wireless communications by exploiting the physical-layer characteristics of wireless channels, which is shown to achieve the perfect secrecy from information-theoretic perspective. The following summarizes the main contributions of this book.

© Springer International Publishing Switzerland 2016
Y. Zou, J. Zhu, *Physical-Layer Security for Cooperative Relay Networks*,
Wireless Networks, DOI 10.1007/978-3-319-31174-6_6

In Chap. 1, we have presented various physical-layer security enhancement techniques, namely the information-theoretic security, artificial noise aided security, security-oriented beamforming, and diversity assisted security. We have also discussed cooperative relaying techniques for wireless networks, where two basic relaying protocols (i.e., the AF and DF) are compared between each other. More specifically, in the AF protocol, a relay just simply retransmits its received noisy signal from the source to destination without any sort of decoding, whereas the DF protocol allows the relay to decode its received signal before retransmitting to the destination. Moreover, several advanced relaying methods, including the orthogonal relaying, non-orthogonal relaying and relay selection have been presented to address the multi-relay scenario. Additionally, we have discussed the use of cooperative relays for improving the wireless physical-layer security.

Next, Chap. 2 investigates the physical-layer security for a wireless network consisting of a source and a destination with the aid of multiple relays, where an eavesdropper is assumed with an intention to tap the confidential source-destination transmission. We have presented the best relay selection scheme for protecting the source transmission against eavesdropping, where only the "best" relay is selected among the multiple relays to assist the source-destination transmission. For comparison purposes, the conventional direct transmission and random relay selection have also been considered as benchmark schemes. Specifically, in the direct transmission scheme, the source just directly transmits to the destination without relying on the relays, whereas the random relay selection simply selects a relay in a random manner for assisting the source transmission. We have derived closed-form expressions of the intercept probability for the conventional direct transmission and random relay selection as well as the proposed relay selection schemes over Rayleigh fading channels. Also, we have analyzed the secrecy diversity of these schemes and show that the proposed relay selection obtains the full secrecy diversity, while the direct transmission and random relay selection methods achieve the secrecy diversity order of only one. It has been shown that the intercept probability of proposed relay selection is better that of the conventional direct transmission and random relay selection approaches.

In Chap. 3, we then examine the joint relay and jammer selection for enhancing the wireless physical-layer security of the source-destination transmission with the help of multiple intermediate nodes in the presence of an eavesdropper. In the proposed joint relay and jammer selection framework, a node is selected among the multiple intermediate nodes to act as the relay for assisting the source-destination transmission. Meanwhile, among the remaining intermediate nodes, another node is chosen and used as the jammer for transmitting an artificial noise to confuse the eavesdropper. For comparison purposes, the conventional pure relay selection and pure jammer selection have been considered as benchmark schemes. Specifically, the pure relay selection scheme just selects an intermediate node to act as the relay, whereas the pure jammer selection only chooses an intermediate node to act as the jammer. We have derived the closed-form expressions of intercept probability for the proposed joint relay and jammer selection as well as the conventional pure relay selection and pure jammer selection schemes. Numerical results have shown that the intercept performance of proposed joint relay and jammer selection scheme is significantly better than that of the conventional pure relay selection and pure jammer selection methods.

In addition, Chap. 4 explores the security-reliability tradeoff (SRT) for a wireless network comprised of a source and a destination with the aid of multiple relays in the presence of multiple eavesdroppers. We have presented two relay selection schemes, namely the single-relay selection (SRS) and multi-relay selection (MRS) to improve the security and reliability of the transmission from the source to destination. More specifically, in the SRS scheme, only the single "best" relay is selected to participate in assisting the source-destination transmission. By contrast, the MRS scheme enables multiple relays to simultaneously forward the source transmission to the destination. We have analyzed the SRT performance for the conventional direct transmission as well as the proposed SRS and MRS schemes, where the security and reliability are measured by deriving the intercept probability encountered at the eavesdroppers and the outage probability experienced by the legitimate destination, respectively. Numerical results have demonstrated that for all the direct transmission, SRS and MRS schemes, the intercept probability of the source-destination transmission decreases with an increasing outage probability and vice versus, showing the tradeoff between the security and reliability. It has also been shown that both the SRS and MRS schemes outperform the direct transmission in terms of their SRTs, meaning that the wireless security and reliability can be simultaneously improved by exploiting the relay selection. Additionally, as the number of eavesdroppers increases, the SRT of wireless communications degrades. By contrast, increasing the number of relays can significantly enhance the wireless SRTs of the SRS and MRS schemes and moreover, the MRS performs better the SRS in terms of their SRTs.

Finally, Chap. 5 re-examines the joint relay and jammer selection for improving the SRT performance of wireless communications with the help of multiple relay nodes and friendly jammers. Specifically, a relay is selected among multiple relay candidates to help a source transmit its message to a destination for the sake of enhancing the wireless reliability. Meanwhile, a friendly jammer is chosen to improve the physical-layer security through emitting the artificial noise for confusing and preventing an eavesdropper from decoding the source transmission. For comparison purposes, we also consider the conventional pure relay selection and pure jammer selection as benchmark schemes. We carry out the SRT analysis for the conventional pure relay selection and pure jammer selection as well as the proposed joint relay and jammer selection in terms of deriving their closed-form intercept probability and outage probability over Rayleigh fading channels. Numerical results demonstrate that the SRT performance of the proposed joint relay and jammer selection is strictly better than that of the conventional pure relay selection and pure jammer selection. It is also shown that with an increasing number of relays and jammers, the SRT of the joint relay and jammer selection is enhanced significantly, implying that exploiting the joint relay and jammer selection is capable of improving both the wireless security and reliability concurrently.

6.2 Future Work

This section presents a range of open challenges and future trends for wireless physical-layer security. As discussed above, extensive research efforts have been devoted to enhancing the wireless security against eavesdropping, however various challenging issues still remain open at the time of writing. In what follows, we present several interesting research topics to be explored in the future.

First, most of existing physical-layer security work only addresses the eaves-dropping attack but without jointly considering different types of wireless attacks e.g. a mixed wireless attack scenario comprised of the eavesdropping attack and denial-of-service (DoS) attack. To be specific, the eavesdropping attack is to tap the confidential transmission, while the DoS attack attempts to emit an interference signal for disrupting the legitimate transmission. It would be of interest to explore how to simultaneously defend against multiple different types of wireless attacks, referred to as a mixed wireless attack. In order to protect the wireless transmission against a mixed attacker e.g. including both the eavesdropping and DoS behavior, we should not only maximize the secrecy capacity of wireless transmission for preventing the information leakage, but also minimize the interference impact caused by the DoS attack on the legitimate transmission. As a consequence, the CSIs of the main channel (from the source to legitimate destination), wiretap channel (from the source to eavesdropping attack) and interference channel (from the DoS attack to legitimate destination) may be required for performing the design of security defense mechanisms against the mixed attack. It would be important to explore the security defense design in different scenarios assuming the availability of the full or partial knowledge of the CSIs of the main channel, wiretap channel and interference channel.

Second, the security, reliability, and throughput are always the driving factors for the evolution of wireless communications systems. Presently, the security, reliability, and throughput enhancement mechanisms are designed individually and optimized separately. However, this is not an effective way, since these three driving factors are coupled and affect each other. For instance, increasing the transmit power can effectively enhance the reliability and throughput of the main channel from the source to destination. On the other hand, it also leads to the fact that an improved signal reception quality would be achieved at the eavesdropper, resulting in an increase of the capacity of the wiretap channel from the source to eavesdropper. It can be observed that with an increasing transmit power, the reliability and throughput improvement is obtained at the cost of the security degradation. To this end, it is necessary to explore the joint security-reliability-throughput optimization for achieving the secure, reliable, and fast communications. More specifically, the joint security-reliability-throughput optimization may be targeted on maximizing the security performance under the reliability and throughput constraints. Similarly, the joint optimization problem may be formulated to maximize the throughput with the target security and reliability requirements. Additionally, we may also adopt the reliability as the optimization objective under the wireless security and throughput constraints.

Third, the cross-layer design is largely ignored in the open literature on wireless physical-layer security, which is to be investigated in the future. Conventionally, the layered protocol architecture has been adopted in wireless networks, which is comprised of the physical layer, medium access control (MAC) layer, network layer, transport layer, and application layer. These protocol layers are typically protected separately for meeting the strict security requirements, including the authenticity, integrity, and confidentiality. However, the layered security mechanism is cost-ineffective, since each protocol layer would introduce extra computational complexity and latency. For instance, in order to meet the authenticity requirement, existing wireless networks adopt multiple authentication approaches simultaneously at different layers, including the MAC authentication, network authentication, and transport layer authentication. Although the use of multiple separate authentication methods at different protocol layers enhances the security level, it comes at the cost of high security overhead. Therefore, it will be of importance to explore the cost-effective security for wireless networks to improve the wireless security at a reduced expense of the security overhead (e.g., the computational complexity and latency).

Fourth, the physical-layer security of wireless communications through cooperative relays can be significantly enhanced in terms of decreasing the intercept probability, especially with an increasing number of relays. It is worth mentioning that in previous chapters, the cooperative relays are assumed to be trustworthy without tapping the legitimate wireless transmissions. However, the cooperative relays may be captured and compromised by an adversary. In this case, the relays become untrustworthy and may be potential threats of destroying the information confidentiality. It is thus necessary to explore the detection and prevention of untrustworthy relays, especially under practical constraints in terms of the affordable computational complexity and latency. Also, it is of interest to investigate whether the use of untrustworthy relays is still beneficial in terms of improving wireless physical-layer security. Moreover, it is highly desirable to examine the employment of cooperative relays for assisting the legitimate transmission, while assuring the legitimate transmission completely confidential to the relays.

Additionally, there are various emerging wireless physical-layer security enhancement approaches, including the beamforming, artificial noise, and cooperative relaying. Presently, their security benefits have only been shown in theoretical studies relying on some idealized simplifying assumptions, e.g., the perfect CSI knowledge is typically assumed in the open literature. For example, in the beamforming aided security approach, the accurate CSI of the main channel (spanning from the source to destination) is needed for the beamforming design, so that the signal received at the destination experiences constructive interference, while the eavesdropper encounters destructive interference. However, it is impossible to obtain the perfect CSI in practice, since estimation errors always exist when estimating the CSI, no matter what channel estimation methods are employed. With the inaccurate CSI based beamforming, the signal received at the destination may even experience destructive interference, which would lead to a significant security performance degradation. It remains unknown whether the beamforming aided security method is still attractive when the imperfect CSI is considered. Moreover, it is necessary to conduct field experiments for different security enhancement approaches in practical wireless communications systems for the sake of providing general guidelines in terms of their security benefits and corresponding costs.

Printed in the United States
By Bookmasters